Iterative Learning Control for Deterministic Systems

Kevin L. Moore

Iterative Learning Control for Deterministic Systems

With 35 Figures

Springer-Verlag

London Berlin Heidelberg New York
Paris Tokyo Hong Kong
Barcelona Budapest

Kevin L. Moore
College of Engineering, Idaho State University, Campus Box
8060, Pocatello, Idaho 83209-8060, USA

Cover illustration: Appendix B, Figure 2. Feedforward neural network
with two hidden layers.

British Library Cataloguing in Publication Data
Moore, Kevin L.
   Iterative Learning Control for
   Deterministic Systems. – (Advances in
   Industrial Control Series)
   I. Title   II. Series
   006.3

Library of Congress Cataloging-in-Publication Data
Moore, Kevin L., 1960–
   Iterative learning control for deterministic systems/Kevin L.
Moore.
      p.   cm. – (Advances in industrial control)
   Includes bibliographical references and index.
   ISBN-13: 978-1-4471-1914-2   e-ISBN-13: 978-1-4471-1912-8
   DOI: 10.1007/978-1-4471-1912-8

   1. Intelligent control systems.   2. Neural networks (Computer
science)   I. Title.   II. Series.
TJ217.5.M66   1992   92-15936
670.42'75–dc20   CIP

Softcover reprint of the hardcover 1st edition 1993

Typesetting: Camera ready by author
69/3830–543210   Printed on acid-free paper

*This monograph is dedicated to
my wife Tamra
and our chidren, Joshua and Julia*

SERIES EDITOR'S FOREWORD

The series *Advances in Industrial Control* aims to report and encourage technology transfer in control engineering. The rapid development of control technology impacts all areas of the control discipline. New theory, new controllers, actuators, sensors, new industrial processes, computing methods, new applications, new philosophies, ..., new challenges. Much of this development work resides in industrial reports, feasibility study papers and the reports of advanced collaborative projects. The series offers an opportunity for researchers to present an extended exposition of such new work in all aspects of industrial control for wider and rapid dissemination.

Kevin Moore's text investigates the design of algorithms which enable a process or piece of equipment to execute the same operation repeatedly. Production lines in many manufacturing industries contain examples of such routines, often involving the use of robotic manipulators. Paint spraying and spot-welding in the car industry are two widely cited examples. However, a little thought shows that almost all sequential material processing production lines comprise repeated processes. Thus, it is not too surprising that researchers have been trying to develop algorithms which learn from the performance of a previous processing run.

As a contribution to this economically important field, Kevin Moore reports his recent work and experiences with iterative learning algorithms. As the monograph develops, links with adaptive control and other learning methods begin to appear, in particular neural networks. Moore's enthusiasm for the neural network development is evidenced by a useful tutorial introduction to this rapidly growing area which is also included in this useful Series Volume.

M. J. Grimble and M. A. Johnson
Industrial Control Centre
Glasgow, Scotland, UK

AUTHOR'S PREFACE

An important problem in the development of advanced industrial control systems is the design of controllers to meet transient response performance requirements. Iterative learning control is an approach to this problem for systems or processes that operate repetitively over a fixed time interval. In this research monograph we consider the analysis and design of learning control systems. The volume begins with an introduction to the concept of learning control and a comprehensive literature review. Then, a complete and unifying analysis of the learning control problem for deterministic linear time-invariant (LTI) systems is given. This analysis uses a system-theoretic approach and offers insight into the nature of the solution of the learning control problem. In addition to the analysis, several design methods are given for LTI learning control, including a technique based on parameter estimation and a one-step learning control algorithm for finite-horizon problems. The monograph also considers learning control for deterministic nonlinear systems. A time-varying learning controller is presented that can be applied to a class of nonlinear systems that includes the models of typical robotic manipulators. The application of artificial neural networks to the learning control problem is also discussed. Three specific ways to use neural nets for learning control are described: two methods that use backpropagation training and one method based on reinforcement learning. Finally, an expanded appendix that provides a tutorial on artificial neural networks is included. The monograph should prove useful to graduate students and researchers interested in the problem of iterative learning control and artificial neural networks. It will also be valuable to engineers interested in developing advanced control systems for industrial control and automation applications.

It is a pleasure to acknowledge those who have helped make this monograph possible. Special appreciation is extended to my Ph.D. dissertation advisors: Shankar P. Bhattacharyya of Texas A&M University and Mohammed Dahleh of the University of California at Santa Barbara. "Dr. B" has been my most important mentor and has strongly influenced my perspective on control theory. He also introduced me to the problem of iterative learning control and later encouraged me to apply artificial neural networks to the problem. Mohammed provided me with many helpful ideas and suggestions regarding the technical content of my research. Appreciation is also extended to D. Subbaram Naidu of Idaho State University and to Herschel Smartt, Carolyn Einerson, and John Johnson of the Idaho National Engineering Laboratory, for many stimulating discussions regarding the application of neural nets to control problems. I am also grateful to V. "Hary" Charyulu, Dean of Engineering at Idaho State University, for providing an atmosphere that allowed me to concentrate on the monograph and its contents. Acknowledgement is given to Mark Waddoups for assisting with some of the computer simulations presented in Chapter 7. Acknowledgement is also given to the U.S. Air Force for support under the 1990 Summer Faculty Fellowship program and to the U.S. Department of Energy for support under the 1991 AWU-DOE Faculty Fellowship program. I would also like to thank the Series Editors M.J. Grimble and M.A. Johnson for their careful reading of the manuscript, Springer Engineering Editor Nicholas Pinfield for his support throughout the publication process, and Lynda Mangiavacchi and the staff at Springer-Verlag for their expert assistance in all matters related to the camera-ready copy of the manuscript.

It is impossible to completely express my appreciation and gratitude to my family. They have provided me with all the support I could have asked for during the time this work was being completed. I offer my thanks for their love, their encouragement, and their sacrifices, without which I could not have produced this monograph.

Pocatello, Idaho   Kevin L. Moore
                                                                        April 1992

ACKNOWLEDGEMENTS

The ideas on iterative learning control presented in this monograph were developed over a period of several years and have previously appeared in a number of different publications, with varying degrees of completeness. The purpose of the current work is to provide a complete and unified presentation of the subject by bringing these ideas together in a single publication. To achieve this goal, it has been necessary at times to reuse some material that we reported in earlier papers. Although in most instances such material has been modified and rewritten for the monograph, permission from the following publishers is acknowledged.

We acknowledge the permission of the International Federation of Automatic Control to reproduce portions of the following paper.

> Kevin L. Moore, S. P. Bhattacharyya, and Mohammed Dahleh, "Some Results in Iterative Learning", in *Automatic Control, Vol. IV, Proceedings of the 11th IFAC World Congress*, U. Jaaksoo and V. I. Utkin (Editors), Tallin, Estonia, Pergamon Press, Oxford, pp. 165–170, August 1990.

Springer-Verlag, Berlin is acknowledged for permission to reuse portions of the following book chapter.

> Kevin L. Moore, S. P. Bhattacharyya, and Mohammed Dahleh, "Learning Control for Robotics", Chapter 22 (pp. 240–251) of *Advances in Computing and Control*, vol. 130 in the series Lecture Notes in Control and Information Sciences, Springer-Verlag, Berlin, 1989.

Acknowledgement is given to the Institute of Electrical and Electronic Engineers for permission to reproduce parts of the following papers.

> Kevin L. Moore, Mohammed Dahleh, and S. P. Bhattacharyya, "Iterative Learning for Trajectory Control", in *Proceedings of the 1989 Conference on Decision and Control*, Tampa, Florida, pp. 860–865, December 1989.

Kevin L. Moore, Mohammed Dahleh, and S. P. Bhattacharyya, "Adaptive Gain Adjustment for a Learning Control Method for Robotics", in *Proceedings of the 1990 IEEE International Conference on Robotics and Automation*, Cincinnati, Ohio, pp. 2095–2099, May 1990.

Kevin L. Moore, S. P. Bhattacharyya, and Mohammed Dahleh, "Arbitrary Pole and Zero Assignment with N-Delay Input Control Using Stable Controllers", in *Proceedings of the 1989 Conference on Decision and Control*, Tampa, Florida, pp. 1253–1258, December 1989.

Kevin L. Moore, "Artificial Neural Networks", *IEEE Potentials*, vol. 11, No. 1, pp. 23–28, February 1992.

We acknowledge permission from the American Automatic Control Council to reproduce portions of the following paper.

Kevin L. Moore, "A Reinforcement-Learning Neural Network for the Control of Nonlinear Systems", in *Proceedings of the 1991 American Control Conference*, Boston, Mass., pp. 21–22, June 1991.

CONTENTS

LIST OF FIGURES

CHAPTER 1

INTRODUCTION TO THE MONOGRAPH

Recently, increased attention has been given in the media and trade journals to the relationship between a country's performance in the industrial sector and its success in the international marketplace. It has been generally observed that future excellence in manufacturing and industrial automation is essential to improving quality and competitiveness in the industrial sector. An important factor in providing the technological base necessary to achieve this excellence is the development of advanced industrial control systems. As more sophisticated control strategies are developed for practical systems (such as robotic manipulators, motion control systems, and process control systems), it will be possible for the manufacturing and industrial control industries to reduce production costs and increase their competitiveness.

There are a number of issues to address in designing advanced control systems. One problem of particular interest is the design of controllers to guarantee that a desired trajectory or motion will be executed by the system with acceptable accuracy. This problem is of obvious importance in the area of robotics and automation, but is also important in other applications, such as those related to motion control. A new approach to solving this problem is *iterative learning control*. Iterative learning control is a technique for improving the performance of systems or processes that operate repetitively over a fixed time interval. In practice this defines a broad class of systems to which the technique can be applied.

This research monograph considers the analysis and design of iterative learning control systems. We present a complete and unifying theory of iterative learning control for linear systems and we also consider the learning control problem for nonlinear systems, such as robotic manipulators. We also incorporate ideas from the emerging technology of artificial neural networks and show how they can be applied to the learning control problem. Throughout the monograph we restrict our discussion to deterministic dynamical systems. A study of learning control from the perspective of stochastic control and automata theory is a separate subject, which we will not consider in this volume (although we do present one result that uses a stochastic automaton as a learning controller for a deterministic system). The monograph directly contributes to the technological base necessary to realize advanced industrial control systems, by developing and demonstrating advanced control algorithms that can be applied to robotic manipulators, motion control problems, and other problems that involve repetitive operation of the system.

In this chapter we first describe the background that motivated the present study of iterative learning control. We then outline the organization of the monograph.

1.1 BACKGROUND AND MOTIVATION: TRANSIENT RESPONSE CONTROL

We were initially led to the study of iterative learning control by our interest in the transient response control problem. Consider Figure 1.1, which shows the configuration of a unity feedback control system. Here $\mathcal{P}$ represents the system to be controlled, usually called the plant or process, and $\mathcal{C}$ denotes the controller. The

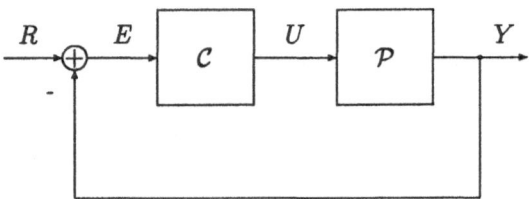

Figure 1.1: Unity feedback control system.

control system operates by measuring Y, the output of the plant, and comparing it to the reference input R. The error E is then input to the controller, which computes an appropriate actuating signal U. This signal is then used to drive the plant. A typical design problem is to specify a controller for a given plant, so that the closed-loop system is stable and the output tracks a desired trajectory with a prescribed set of transient performance characteristics (such as overshoot, settling-time, rise-time, steady-state error, etc.). When the plant is a linear time-invariant (LTI) system there are a variety of ways to choose a controller that

2

ensures closed-loop stability, achieves acceptable steady-state error performance with respect to reference and disturbance inputs, and provides for pole placement or the minimization of a cost functional [1]. However, successful design for specific transient response performance is more difficult. One reason is that there is no analytic relationship between desired transient response characteristics and pole/zero locations for systems of order higher than two. Conventional optimal control methodologies suffer from the same limitations. For instance, it is not always clear how a given transient response specification can be translated into an appropriate cost functional. Also, even if a desired transfer function between the input and output is specified, it may not be attainable using an LTI controller [2]. Thus, in some cases a conventional controller design may never be able to achieve a specified transient response. For the case of nonlinear or time-varying plants the problem is worse, with few available results. Additionally, the problem of robustness makes it even harder to realize a desired transient response. In most cases, actual transient performance during system operation typically will not match the designed performance, due to modelling uncertainty, or drift and aging of components. For these reasons, there is a need for transient response design strategies that will (i) overcome the analytic uncertainty inherent in conventional controller design and, (ii) account for modelling uncertainties that are exhibited during system operation. It was this problem that originally attracted our attention. This in turn brought us to the study of iterative learning control. For completeness, the following paragraphs outline the path we followed, from our initial interest in transient response design to the study of iterative learning control.

First, rather than specifying a design to meet a complete transient response requirement, we concentrated on a single performance criterion: overshoot. In general it is not known where the closed-loop system poles and zeros should be placed to yield small overshoot. However, we can simplify the problem by asking where we should place the free zeros of the system, assuming the poles are already specified (by bandwidth considerations, for instance). Motivated by a contribution from Dahleh and Pearson on the design of optimal overshoot controllers [3], we formulated a design technique for single-input, single-output systems which specifies the minimum overshoot controller when the controller order is fixed and the closed-loop system poles have a prescribed location [4]. The effect of this design technique is to find the zeros of the controller that result in the smallest possible overshoot for a given set of closed-loop system poles (assuming the controller is

3

of pole placement order or higher).

However, a fundamental limitation of this design technique is that the plant zeros cannot be changed. Recently, multirate digital control schemes have been proposed to overcome this limitation. A study of these ideas led us to some initially promising results. Specifically, starting from a paper by Mita and Chida [5], we developed some results that allow us to arbitrarily assign the complete set of closed-loop poles and zeros in a multirate digital control system using output feedback [6]. In this technique the input is changed more often than the output is sampled. We call this approach N-delay input control, and our results showed that we can completely specify the transient performance at the output sample points in a digital control system. Unfortunately, we found that in systems designed using this approach, the intersample behavior of the system was degraded. Because our analysis of this effect indicated that the apparent promise of the approach cannot be realized, we began to think about other possible system configurations for dealing with the transient design problem.

The two techniques described above both deal with the specification of the parameters of a feedback controller that both stabilizes the system and forces a desired response. Another approach to control system design is to use a feedback controller to stabilize the closed-loop system and then design for transient response behavior with a prefilter or a feedforward controller. For this we require a two degree-of-freedom configuration. One such configuration is shown in Figure 1.2. With this system structure, the signal R in Figure 1.1 has become $R = \mathcal{H}_p Y_d$, where $\mathcal{H}_p$ is the prefilter and Y_d is the desired response. The prefilter $\mathcal{H}_p$ gives us an extra design parameter. Another related configuration is shown in Figure 1.3. In this figure, the output of the block $\mathcal{H}_p$ is "fed forward" into the loop.

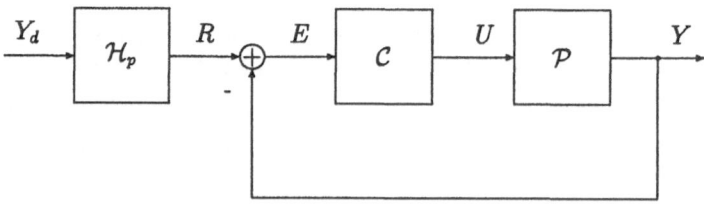

Figure 1.2: Control system with a prefilter.

This is called a feedforward configuration and is also a two degree-of-freedom system. It is possible to state the conditions under which Figure 1.2 and Figure 1.3 are equivalent. In either case, however, the idea is that $\mathcal{C}$ is used to stabilize the system (and perhaps provide for steady-state tracking) and the prefilter or feedforward block $\mathcal{H}_p$ is designed to meet transient performance requirements. But again, we are led to the question of exactly how to choose the best prefilter to satisfy transient response requirements.

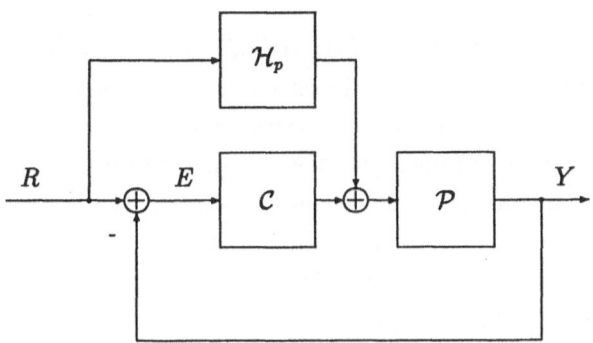

Figure 1.3: Control system with a feedforward block.

It was the question of how to design an optimal prefilter that first led us to the study of iterative learning control. At its most fundamental level, learning control can be thought of as an approach to finding the output of an optimal prefilter for a given system. However, as opposed to an *a priori*, analytical design technique, learning control is an iterative technique, which "learns" the output of the optimal prefilter, using actual input/output data obtained in real-time during system operation. In the sequel we will describe the approach of iterative learning control. We will show how learning controllers can be designed and how they can be used as a viable solution to the problem of transient response control.

1.2   ORGANIZATION OF THE MONOGRAPH

In this monograph we present a number of results related to the analysis and design of iterative learning control systems. The presentation of the results is organized as follows. In Chapter 2 we describe the concept of learning control. We then present a comprehensive literature review, which summarizes the current status of the problem. This is followed by a very general formulation of the learning control problem, using operator-theoretic notation. This formulation is based on a contraction mapping approach and includes as special cases most of the approaches to learning control given in the literature.

Chapters 3, 4, and 5 present a complete analysis of the learning control problem for linear time-invariant plants with a proposed class of causal, linear time-invariant learning controllers. There are three main contributions of this analysis. First, in Chapter 3, we develop general convergence conditions for the proposed class of LTI learning controllers. These conditions offer insight into the nature of the solution of the learning control problem and also provide a measure of the best possible performance attainable via learning control. The implication of this analysis is that, for an LTI plant, the best possible LTI learning controller generates the output of the closest approximate inverse of the system for a desired output. This implies that, in the presence of full plant knowledge, learning control can do no better than *a priori* design. This fact is intuitive and is perhaps implicitly assumed in the literature but is not always explicitly stated. Second, in Chapter 4, a learning control scheme based on parameter estimation is proposed for use with unknown LTI plants. This method is compared to traditional adaptive control schemes. Third, in Chapter 5, a study of finite-horizon learning control problems is given that shows how to develop a one-step learning controller with "memory". That is, the learning controller not only produces the output of the best approximation to an inverse system after only one trial, but it also learns the parameters of such an inverse over the interval of interest. We also present a learning control strategy for non-minimum phase systems that is based on multirate sampling techniques (2-delay input control). This scheme converges in two steps. Finally, we discuss some extensions of these ideas for time-varying systems.

We noted above that the essential effect of a learning control scheme for LTI systems is to iteratively produce the output of the closest possible inverse of the system. This implies that for LTI problems learning control may offer no real advantage over conventional design techniques. However, such a conclusion does not follow when we consider nonlinear or time-varying plants. In Chapter 6 we discuss learning control for nonlinear systems. Then, to illustrate the potential usefulness of learning control for nonlinear systems, we present a time-varying learning controller for a class of nonlinear systems that includes typical models of jointed robotic manipulators. The contribution of this chapter is a learning control scheme that improves on a technique given by Bondi, et al. [7]. The method is demonstrated with a simulation example.

The observation that the real effectiveness of learning control may be found in its application to nonlinear or time-varying systems leads us to consider the possibility of using some type of nonlinear structure for the learning controller itself. Recently there has been a renewed interest in the study of artificial neural networks. Artificial neural networks are models for computing that are motivated by biological systems. These models exhibit self-organizing and learning properties that are of interest in the context of artificial intelligence and machine learning. From a control systems point of view, an artificial neural network can be thought of as a large-scale nonlinear system. Thus, it seems natural to consider the possibility of developing learning controllers based on neural networks.

In Chapter 7 of the monograph we consider the use of artificial neural networks for learning control. We present three different ways of using an artificial neural network structure for the learning controller. In all three cases we exploit the ability of a neural network to implement and learn nonlinear mappings from input to output space. Each method is developed and illustrated with examples.

Chapter 8 concludes the monograph with a summary of our contributions and a discussion of future research directions. Additionally, two appendices are included. Appendix A describes the essential results needed to understand the learning control method for non-minimum phase systems described in Chapter 5. Appendix B provides an introduction to the field of artificial neural networks to support the discussion of neural net learning controllers in Chapter 7. This expanded appendix can be used as a tutorial on neural nets for those readers who have had little or no exposure to this area.

CHAPTER 2

ITERATIVE LEARNING CONTROL: AN OVERVIEW

In this chapter we give an overview of the field of iterative learning control. We begin with an introduction to the concept of learning control. Then we summarize the past work that has been reported in the literature. We conclude the chapter with a general formulation and statement of the learning control problem.

2.1 INTRODUCTION

Iterative learning control is an approach to improving the transient response performance of systems that operate repetitively over a fixed time interval. It is useful for problems in which a system must be able to follow different types of inputs, in the face of design or modelling uncertainty.

The concept of learning control was motivated by observing the behavior of systems that operate repetitively. Consider an antenna servomechanism for a radar system. A typical motion executed in scan mode is shown in Figure 2.1. The antenna line-of-sight is moved across in azimuth, stepped up in elevation, slewed back the opposite direction in azimuth, and finally stepped back down in elevation, ending the trajectory at the original starting point. This motion is then repeated over and over again. Now, suppose that the controller design for this antenna yielded a ten percent overshoot at each corner of the scan pattern, with a five millisecond settling time. Then, each time the scan was executed, the system would exhibit these same values of overshoot and settling time. A natural question to ask about this would be the following: if we know that we obtain a ten percent overshoot and five millisecond settling time at the same point in the trajectory each time the motion is executed, then why not use this knowledge to improve the trajectory response? That is, why not measure the performance error each time we perform the scan pattern, and then change our control strategy the next time the trajectory is commanded. This is what we do in an iterative learning control scheme. However, the way we change our control strategy is by

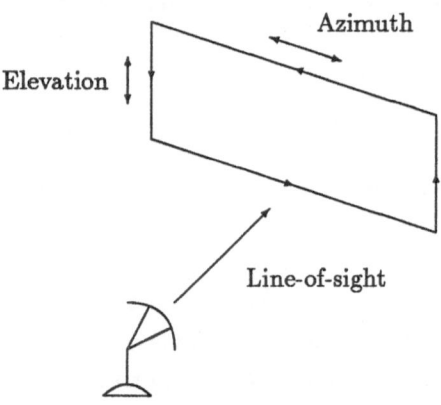

Figure 2.1: Antenna servomechanism scan pattern.

changing the commanded reference signal, as opposed to changing the controller itself, as in a conventional adaptive control paradigm.

Another motivation for the approach of iterative learning control arises from problems in which a system must have the capability to accurately respond to several different types of inputs. An example of such a system is a robot arm in a flexible manufacturing assembly line. A robot arm in this situation must execute a given trajectory many times, but may be configured to perform several different tasks, depending upon production needs. For instance, a robotic manipulator may have the capability to be fitted with several different end-effectors. Perhaps one week the manipulator is employed painting white-walls on automobile tires. Later, when production requirements change, the robot might be assigned to weld a horizontal seam on the side of door panels. In either case, the motion of the manipulator is repetitive, but the motions required for the two tasks are quite different and hence the control action for each case will have to be different. In many conventional control schemes the controller itself may require a different structure to force steady-state tracking of different types of inputs (to say nothing of transient performance). For instance, to track a circular (sinusoidal) input command for painting whitewalls on tires, the internal model principle of classical control tells us that the controller denominator polynomial will have to contain terms such as $(s^2 + \omega^2)$. However, to weld a seam on a door by following a ramp input the open loop will need to contain a double integrator. In order to respond to both types of inputs we might design a single controller that can handle either contingency, or we might design two controllers, with some type of logic for switching between the two designs. Alternately, we could use the method of learning control to learn the correct reference signal for each desired trajectory.

To explain the idea of modifying the input to the system by training, consider one final example. A child is throwing stones at a log submerged in a stream. At first, the stones miss the target because the refractive index of the water gives a misleading sense of its location in the water. Soon, however, the child learns to compensate for this effect by throwing the stones slightly off from the perceived mark. Notice that this is not done by changing any fundamental structure of the sensory-motor control system, which still observes the log to be at the wrong place. Instead, the child changes the command to the muscles of his arm, telling them to throw at the different mark. In control system language we might express it in the following way. If we command the system to follow a step input, but the performance is unsatisfactory, then we could try to compensate in the same way as the stone thrower: tell the system to follow a signal other than a step. In this way we might "fool" the system into the desired performance, much in the same way that our stone thrower fooled his sensory-motor system. The key, of course, is to determine how to define this other signal that will drive the system to the desired trajectory. Iterative learning control is an approach to finding such an input signal.

The concept of iterative learning for generating the optimal input to a system was first introduced by Uchiyama [8] (in Japanese). The idea was later developed by Arimoto and his co-workers [9]–[21]. Figure 2.2 illustrates the basic idea. Each time the system operates, its input and output signals ($u_k(t)$ and $y_k(t)$, respectively) are stored in memory (some type of memory device is implicitly assumed in the block of Figure 2.2 labelled "Learning Controller"). The learning control algorithm then evaluates the performance error $e_k(t) = y_d(t) - y_k(t)$, where $y_d(t)$ is the desired output of the system. Based on the error signal, the learning controller then computes a new input signal $u_{k+1}(t)$, which is stored for use on the next trial (by "the next trial" we mean the next time the system is operated). The next input command is chosen in such a way as to guarantee that the performance error will be reduced on the next trial.

Notice that we have defined our signals with two variables, k and t. The trial is indexed with the integer k, while time is described by the variable t, which may be continuous or discrete. When we say we store $u_k(t)$ we mean that we

11

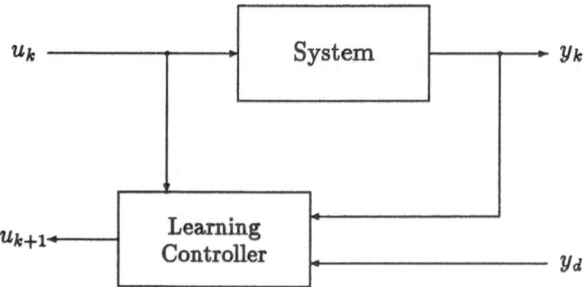

Figure 2.2: Learning control configuration.

have stored the time signal $u_k(t)$ (or a finite part of it) that was used to drive the system during the kth trial or repetition of the system trajectory. Also, notice that we usually assume that all the initial conditions of the system are reset to the same value at the beginning of each new trial. This is a reasonable assumption, which is typically satisfied in practice.

The important task in the design of a learning controller is to find an algorithm for generating the next input in such a way that the performance error is reduced on successive trials. This is usually quantified by saying that the error should converge, with convergence measured in the sense of some norm. Also, we would like the learning control algorithm to cause convergence of the error without knowing the model of the plant under control (or, at least, it should require minimal knowledge of the system parameters). Further, the algorithm should be independent of the functional form of the desired response $y_d(t)$. Thus, the learning controller would "learn" the best possible control signal for a particular desired output trajectory without the need to reconfigure the algorithm. Then, if a new desired trajectory is introduced, the learning controller would simply "learn" the new optimal input, without changing any of its own algorithms, by using input-output data obtained from actual operation.

Note that iterative learning control differs from both optimal control and adaptive control. In optimal control we conduct an *a priori* design, based on a model of the system. If the plant changes relative to the model then the optimal controller will cease to be optimal (although adaptive LQR/LQG control algorithms can be used). On the other hand, if the plant changes in a learning control scheme, the learning controller adapts by adjusting the input for the next trial, based on the measured performance error of the current trial. However, iterative learning control is different from conventional adaptive control (although in a very general sense of the word we can describe learning control as adaptive). Most adaptive control schemes are on-line algorithms that adjust the controller's parameters until a steady-state equilibrium is reached. In a learning control scheme, it is the commanded reference input that is varied (in an off-line fashion), at the end of each trial or repetition of the system.

2.2 LITERATURE REVIEW

In this section we will describe our understanding of the state-of-the-art in the field of iterative learning control. When attempting such a review, it is inevitable that some work will be inadvertently omitted. However, the large number of conferences and journals available today make it impossible to be aware of every contribution on a given subject. This is especially true in the field of learning control, which is discussed in a wide range of contexts, from robotics, to artificial intelligence, to classical control, to neural networks. Nonetheless, we will attempt to identify the important contributions in the field and to provide a basic historical context for the study of iterative learning control.

Before discussing the literature, we first comment on the phrase "learning control." In this phrase, the word "learning" is often the cause of misunderstanding. This is especially true given the current interest in artificial neural networks. "Learning" is a broad concept, which means many things to many people (in one discussion, a colleague thought an upcoming lecture on the topic of "learning control" would deal with a new pedagogical approach to control system education!). In a general sense, learning refers to the action of a system to adapt and change its behavior based on input/output observations. In the cybernetics literature the term learning has been used to describe the ability of a system to respond to changes in its environment. Many systems have this ability, including: adaptive control systems, any type of artificial neural network equipped with a weight update algorithm, and more complex systems developed in the theories of psychology (such as hierarchical models of "schemas," used to explain memory and its development). Hence, when using the term learning control, we must be careful to restrict ourselves to the meaning as defined in the previous section. That is, although learning is taking place, in the general sense of the word, we assume the

13

structured architecture shown in Figure 2.2, where the concern is with iteratively generating a sequence of input trajectories, u_{k+1}.

Further confusion may also arise when placing the word "learning" together with the word "control." This is because the term "learning control" is not unique in the control literature. Researchers in the fields of adaptive control, stochastic control, cybernetics, and optimal control have all used the term learning control to describe their work. For instance, in the area of adaptive control, see the classic works on learning and adaptation by Tsypkin [22,23], the survey paper by Fu [24], or an ACC symposium reprint [25]. One point about most of these references, however, is that they refer to learning in the sense of adapting or changing controller parameters on-line, as opposed to the off-line learning described in the previous section. Early references to learning also arose in the context of stochastic control and cybernetics. See, for example, Narendra [26]. In one reference from the area of stochastic control, the distinction between off-line learning and the on-line learning of conventional adaptive control has been pointed out (Sardidis [27]). In this book the term "training" is used to describe the off-line learning control process. The idea of iterative techniques has also been used in the context of optimal control. In [28], Plant used iterative techniques to improve the computation of optimal control solutions. In [29], a type of learning control method was applied by LaCarna and Johnson to adjust the optimal controller feedback gain matrix for the control of an interconnected power system.

Other terms have been introduced in the literature to describe the process of off-line learning, including: "betterment process" [9], "iterative control" [30], and "repetitive control" [31]. (Again we advise caution. The term "repetitive control" has also been used by Hara et al. [32,33], but the context in these papers was the problem of periodic control and not the type of learning control we are considering.) Despite the different names that have appeared, the term that has become standard is "iterative learning control, " coined by Arimoto et al. [11]. In this monograph we will use the phrases learning control and iterative learning control to refer to off-line learning as depicted in Figure 2.2.

As noted above, iterative learning control in the sense of Figure 2.2 was first introduced by Uchiyama [8] in 1978. Because his idea of learning through repeated trials was published in Japanese, it was not widely known in the West. Later the idea was developed by a research group centered around Suguru Arimoto. He and his colleagues began publishing their work in English [9]-[21], which led to increased interest in the idea of iterative learning control, particularly through the middle to late 1980's. Although most of the early work applied to linear plants, results for nonlinear systems were also developed, motivated primarily by models from robotics. In the next few paragraphs we will describe the progress that has been reported in the literature. We will first describe the work of Arimoto and his colleagues. Then we will consider other results dealing with learning control for linear systems. Finally, we discuss the work on learning control for nonlinear systems, including applications to robotics.

The first learning control scheme proposed by Arimoto et al. involved the derivative of the error $e_k(t) = y_d(t) - y_k(t)$ [9]. Specifically, the algorithm had the form

$$u_{k+1} = u_k + \Gamma \dot{e}_k.$$

Here Γ is a constant matrix. Assume that the plant is a linear time-invariant (LTI) system, with signals defined over the interval $(0, t_f)$, and with a state-space description (A,B,C,D). Also assume that CB has full rank. Arimoto et al. showed that if the induced operator norm $\|I - CB\Gamma\|_i$ satisfies

$$\|I - CB\Gamma\|_i < 1,$$

and some initial condition requirements are met, then

$$\lim_{k \to \infty} y_k(t) \to y_d(t).$$

Convergence of the error is in the sense of the λ-norm, which is defined as

$$\|x(t)\|_\lambda = \sup_{0 \le t \le t_f} \{e^{-\lambda t} \max_{1 \le i \le r} |x_i|\},$$

where $x(t)$ is an r-dimensional vector. This norm ensures pointwise convergence over the interval of interest. In this same reference, Arimoto et al. also gave convergence conditions for this type of learning control algorithm when it was applied to time-varying plants and certain nonlinear models encountered in robotics. In subsequent papers (see the bibliography), Arimoto and his co-workers have proposed a variety of different learning control algorithms. The primary differences between the various approaches they have developed is in how the error is utilized in the learning control algorithm. The most general linear algorithm presented is found in [14], where the input is updated according to

$$u_{k+1} = u_k + \Phi e_k + \Gamma \dot{e}_k + \Psi \int e_k dt.$$

15

This algorithm essentially forms a PID-like system for processing the error, while maintaining a linear effect on the past input signal. One note about most of these algorithms is that it is necessary to have knowledge of the input and output maps, B and C, from the state-space description of the system. However, they seem to be quite robust, and Arimoto and his colleagues have been able to demonstrate the effectiveness of their algorithms with actual robots. In more recent work [20,21] their group has considered how to use the knowledge gained from learning several trajectories to improve the rate of learning for a new trajectory.

(As an aside, we would note that Arimoto et al.'s original result is easily extended to the case where CB does not have full rank. Suppose we have a linear system with a pole-zero excess of j (i.e., let j be the smallest integer such that $CA^{j-1}B$ has full rank). Then we can show that the learning algorithm

$$u_{k+1} = u_k + \frac{d^j}{dt^j}e_k$$

will converge if

$$\|I - CA^{j-1}B\Gamma\|_i < 1.$$

Although we state this without proof, it is obvious from the results of the discrete-time, finite-horizon problem discussed in Chapter 5.)

Next we consider some other results related to LTI learning control. Togai and Yamano [34] consider learning control for discrete-time linear systems. They propose the learning control algorithm

$$u_{k+1}(t) = u_k(t) + Ge_k(t+1),$$

where t is an integer variable. Notice that the algorithm looks ahead one step in processing the error, giving a discrete-time equivalent of a derivative (this is acceptable, except at the endpoints, because the processing is done off-line). This makes their learning controller an obvious extension of Arimoto et al.'s original continuous-time scheme. For this update algorithm, the gain G is optimized using gradient methods to minimize the quadratic cost of the error

$$J = \frac{1}{2}e_k^T(i+1)Qe_k(i+1).$$

between successive trials. The authors consider several techniques for choosing G, specifically using the steepest-descent, Gauss-Newton, and Newton-Raphson methods. The first two result in a constant gain G, giving a learning controller with exactly the same form as Arimoto et al. For the Newton-Raphson method the result is a time-varying gain G_k which is different for each trial (although constant on a given trial). In each of these learning control approaches, the only knowledge of the plant needed is the input coupling matrix B from the state-space description.

An approach similar to that of Togai and Yamano is given by Furuta and Yamakita [35]. They present a system-theoretic treatment of the learning control problem. Their update algorithm has the form

$$u_{k+1} = u_k + \epsilon_k T_p^* e_k,$$

where T_p^* is the adjoint operator of the system (a concept from functional analysis) and ϵ_k is a time-varying gain. This gain is computed from the error and the adjoint to provide a steepest-descent minimization of the error at each step of the iteration. Working in a Hilbert space setting, their algorithm provides convergence in the sense of the L_2 norm. However, the method requires complete knowledge of the adjoint of the system, which is equivalent to needing complete knowledge of the system dynamics. To deal with this problem, the authors also provide some singular value conditions that ensure robustness of convergence when there is parameter uncertainty.

In the approaches to learning control mentioned so far, the algorithms all use a unity weighting on the current input. Mita and Kato [30,31] present some results that allow separate weighting of both the current input and the current error. Their algorithm is also distinguished by operating in the frequency domain (actually, the learning control algorithm is executed in the time domain, but the analysis is in the frequency domain). The input update law is defined by

$$U_{k+1}(s) = L(s)[U_k(s) + aE_k(s)].$$

They show conditions under which this type of learning control algorithm will converge in the sense of the L_2 norm (time or frequency). As we note in Chapter 3, this type of algorithm will always produce a non-zero error. Mita and Kato show that this error can be made arbitrarily small and they present a specific design technique for finding the transfer matrix $L(s)$ and the parameter a. However, complete knowledge of the system transfer matrix is necessary to properly choose the filter $L(s)$. We present a generalization of this result in Chapter 3 that indicates explicitly the trade-off between the final error and the parameters of their learning controller.

Other researchers have studied learning control for linear systems. Atkeson and McIntyre [36] have considered learning control schemes similar to Arimoto at al., with application to linear robotic manipulator models. Hideg and Judd [37] have analyzed learning control schemes from a frequency domain perspective, similar to Mita and Kato. In addition, they considered the effect of disturbances on the performance of the learning control system and have also applied their method to linear robotic models (treating the nonlinearities as disturbances). A different approach to learning control is given by Oh, Bien, and Suh [38] for a class of linear time-varying plants. They use a parameter estimator together with an inverse system model to generate the new input for each trial. Their technique is also applied to the learning control of a robotic manipulator. In Chapter 4 we give a related method for LTI systems.

Researchers have also considered the learning control problem for classes of nonlinear systems. Of particular interest to many researchers are the classes of nonlinear systems representative of robotic manipulator models. Arimoto et al.'s original work included a discussion of learning control for robotics [9]. Craig [39]-[41] has independently proposed learning and adaptive schemes similar to those of Arimoto et al., as have Gu and Loh [42]. Harokopos has considered learning from the viewpoint of minimizing a cost functional [43], similar to that of Togai and Yamano, with application to robotics. Bondi, Casalino, and Gambardella [7] give a high-gain feedback, model-reference approach to learning control. It is developed for the nonlinear equations that are typical of robot manipulator models and involves a linear output feedback controller with a linear error feedback learning controller. In Chapter 6 we present an improved version of this scheme that uses an adaptive gain adjustment technique, resulting in a time-varying learning controller. Yamakita and Mita extended their work on linear systems to nonlinear systems in [44] by making use of Gateaux derivatives. In this paper they introduce the term "virtual reference" to describe the optimal input signal derived by the learning controller. They also report a remarkable demonstration of a learning control experiment in which a robot arm iteratively learns to catch a ball in a cup (the Japanese Kendama game). In [45], Bien, Hwang, and Oh extend the parameter estimation results of [38] to handle the problem of robot path control.

Learning control for nonlinear systems has also been considered independent of robotics. A recent paper by Messner, Horowitz, Kao, and Boals considers learn-

18

ing control for nonlinear systems, based on a new method of nonlinear function identification [46]. This method has been applied to learning control for robot manipulators in [47]. Wang [48]-[50] has applied the idea of learning control to the problem of determining the inverse dynamics of an unknown nonlinear system. Hauser [51] has given an analysis of learning control for nonlinear systems that can be modelled as

$$\dot{x} = f(x,t) + B(x,t)u,$$
$$y = g(x,t).$$

Heinzinger, Fenwick, Paden, and Miyazaki [52] have studied the robustness properties of learning controllers for this same type of nonlinear system. Another study on the robustness properties of learning control systems is given by Sugie and Ono [53].

Several researchers have also considered learning control as a special case of learning in general, in the context of intelligent systems [54,55]. Another new approach to learning control is the introduction of ideas from the theory of artificial neural networks [56]. Preliminary indications are that these are promising directions for investigation. We will explore this in more detail in Chapter 7.

2.3 PROBLEM FORMULATION

We will describe our problem in operator-theoretic notation. Let U and Y be normed linear spaces over the reals whose elements are functions taking values in C^r and C^m, respectively. We will assume the same norm is used in each space. Recall that a norm gives a measure of "size" for elements in a linear space (for a review of norms and related concepts, see the introductory chapters in any book on functional analysis or nonlinear systems theory). Let $T : U \mapsto Y$ be a bounded linear operator mapping elements in U to those in Y. We write this as $y = Tu$, where $u \in U$ and $y \in Y$. We also define the induced norm (a measure of "gain") of the linear operator T as

$$\|T\|_i = \sup_{\|x\|=1} \|Tx\|$$

and we denote the composition of two linear operators T_1 and T_2 by $T_1 T_2$.

19

With this notation we will describe a linear system S by

$$y = T_s u,$$

where $u \in U$, $y \in Y$, and the system operator is denoted as T_s. For instance, suppose S is a causal, LTI system that is assumed to be stable and initially at rest. (If we initially have an unstable plant, then we first stabilize it using conventional techniques.) Let S have the state-space description

$$\dot{x} = Ax + Bu,$$
$$y = Cx + Du,$$

where u and y are time functions taking values in R^r and R^m, respectively, and A, B, C, and D are constant matrices of appropriate dimensions. Then we may express $y(t)$ in terms of $u(t)$ and the system matrices as

$$y(t) = C \int_0^t e^{A(t-\tau)} Bu(\tau) d\tau + Du(t).$$

Alternately, assume the system is described in the frequency domain by the input-output relation

$$Y(s) = H(s)U(s),$$

where $H(s)$ is an $m \times r$ transfer matrix and $U(s)$ and $Y(s)$ are functions of the complex variable s, taking values in C^r and C^m, respectively. $U(s)$ and $Y(s)$ are the respective Laplace transforms of the time signals $u(t)$ and $y(t)$ and, for zero initial conditions, we have

$$H(s) = C(sI - A)^{-1} B + D.$$

Each of these descriptions may be expressed by the operator notation $y = T_s u$. For the state-space description T_s is the integral expression given above, while in the frequency domain representation T_s is simply the transfer matrix $H(s)$.

We will also consider nonlinear operators from U to Y. To avoid confusion we will use uppercase letters for linear operators and lowercase letters to denote nonlinear operators. The nonlinear operator $f : U \mapsto Y$ mapping elements in U to those in Y will be written as $y = f(u)$ where $u \in U$ and $y \in Y$. Composition of two nonlinear operators f and g, $f(g(\cdot))$, will be written $fg(\cdot)$.

Now we will present a very general formulation of the problem of learning control. As we have described above, the problem is essentially this: given a system or plant P, we wish to find the optimal input $u^*(t)$, so that if this input is applied to the system, then the resulting output will be "close" to a desired output signal, $y_d(t)$. That is, suppose the system is described by

$$y(t) = f_P(u(t), t).$$

Then we seek an input $u^*(t)$ so that the output signal satisfies

$$y^*(t) = f_P(u^*(t), t) \approx y_d(t)$$

for all time in a specified interval. Learning control is an iterative approach to finding such a $u^*(t)$. The approach of iterative learning control is to generate a sequence of inputs, $u_k(t)$ in such a way that the sequence converges to u^*. This can be stated formally as follows.

PROBLEM : Suppose we are given

(i) a system P, defined by $y_k(t) = f_P(u_k(t), t)$

(ii) a desired response $y_d(t)$, defined on the interval (t_0, t_f).

Then the problem of iterative learning control is to find a system L, defined by $f_L(u, y, y_d, t)$ such that the sequence of inputs produced by the iteration

$$u_{k+1}(t) = f_L(u_k(t), y_k(t), y_d(t), t) = f_L(u_k(t), f_P(u_k(t)), y_d(t), t)$$

converges to a fixed point $u^*(t)$ satisfying

$$\min_{u(t)} \|y_d(t) - f_P(u(t), t)\| = \|y_d(y) - f_P(u^*(t), t)\|,$$

where the norm is defined on the interval (t_0, t_f) and the final time t_f is allowed to go to infinity.

Additional constraints on the problem may include: (i) the initial conditions are reset at the beginning of each trial; and (ii) as little information as possible about the system f_P should be used in designing the system f_L. Notice that this second constraint indicates several levels of problems that we may consider.

First, we can assume full knowledge of the plant to be controlled. This allows us to establish, as a baseline, the nature of the best possible solution to the learning control problem. From this level we can move to more difficult problems, first imposing structure on the plant, but with unknown parameters, and then considering completely unknown systems.

Consider the learning control problem as stated above. In the case of full plant knowledge, the problem is solved if the operator f_P is left invertible and time-invariant. In this case simply let $u^*(t) = f_P^{-1} y_d(t)$. If the system is not invertible it may still be possible to obtain $y(t) = y_d(t)$ for some input $u^*(t)$. This can occur if the plant can be inverted from the direction of the desired output (that is, $y_d(t)$ is in the image of the operator f_P). If $y_d(t)$ is not in the image of f_P then the best we can hope for is to find a $u^*(t)$ that minimizes $\|y_d(t) - y(t)\|$ over all possible inputs $u(t)$. The goal of the learning control algorithm is to iteratively find such a $u^*(t)$.

One other note regards the norms that may be used. As stated, we can consider learning control algorithms defined in either the time domain or in the frequency domain. It simply depends on how one defines the underlying signal spaces and how the associated norms are chosen. However, to practically implement an algorithm designed in the frequency domain it would be necessary to store time signals defined over the entire positive real line. In practice this is not possible, but such schemes will still work, provided we take measurements well past the settling time of the signals. Hence, analysis in the frequency domain will be valid even though our controller will be implemented on a finite interval.

The learning control problem stated here is sufficiently general to include most of the problem formulations presented in the literature. This includes schemes involving time-varying or non-causal learning controllers. However, in this form the problem is so general that it is intractable. To obtain useful results, it is necessary that we place restricting assumptions on the problem: either on the plant, on the learning controller, or both. In the next three chapters we will simplify the problem by restricting our discussion to linear systems. Specifically, we will assume that both the system to be controlled and the learning controller are causal, linear time-invariant systems. These assumptions make it possible to gain insight into the nature of the solution to the learning control problem.

CHAPTER 3

LINEAR TIME-INVARIANT LEARNING CONTROL

Consider again the iterative learning control configuration shown in Figure 3.1. We will now restrict our discussion to the case where the plant is a causal, linear time-invariant (LTI) dynamical system S. We suppose S is represented by the operator T_s, so that $y = T_s u$. We also restrict the learning controller L to be causal and LTI. Although such a plant may not be encountered in practice, we will consider this class as a starting point to gain insight into the problem. As noted in the previous chapter, other assumptions include: (i) the desired response $y_d(t)$ is defined on the interval (t_0, t_f), where t_f may be infinity; and (ii) the initial conditions are reset at the beginning of each trial (although this fact is never used explicitly in any of our proofs, it is used implicitly because we usually suppose that all the initial conditions are zero). To avoid cluttered notation, the time dependence of the various signals will often be suppressed. Additionally, time may be continuous or discrete in our formulation. A final comment is that we are implicitly assuming that the plant T_s is known.

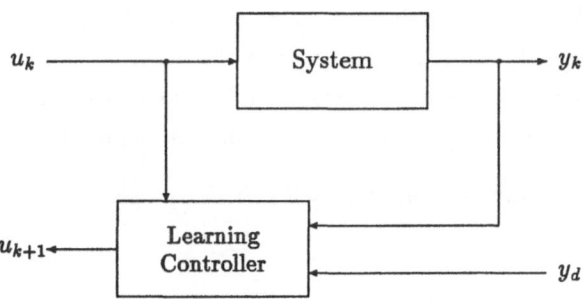

Figure 3.1: Learning control configuration.

Now, the goal of our learning controller is to iteratively produce the signal u^*

that satisfies $y_d = T_s u^*$, or at least makes y_d and $T_s u^*$ as close as possible on the given interval (here the superscript "*" means "optimal"). Specifically, we seek a sequence of inputs u_k with the property that

$$\lim_{k \to \infty} u_k = u^*.$$

Notice that the learning controller operator will be a function of u_k, y_k, and y_d. However, y_k is a function of u_k, because $y_k = T_s u_k$. Hence u_{k+1} really depends only on u_k and the constant signal y_d. As a result, it is natural to look for a fixed point or contraction mapping theorem to use in designing our learning system L. That is, the learning control algorithm should be a contraction operator $T_L(u_k, T_s u_k, y_d)$ whose fixed point is u^*, the optimal input for the system. (As an aside, note that we say a function is Lipschitz if there is a constant ρ such that for all x, y we have $\|f(x) - f(y)\| < \rho \|x - y\|$. If $f(x)$ is Lipschitz, then the iteration $x_{n+1} = f(x_n)$ will converge to a fixed point, denoted x^*, such that $x^* = f(x^*)$. Such a mapping is said to be a contraction.)

To investigate the nature of LTI learning control, we will consider input update laws of the form

$$u_{k+1} = T_u u_k + T_e(y_d - y_k),$$

where T_u and T_e are both linear operators. This is really the most general linear algorithm we might consider because it allows separate weighting of both the current input and the current error. One might argue that a more general form would include separate weightings on the current output y_k and the desired output y_d, rather than a single weighting on the error. However, such an algorithm can be reduced to the one shown through a change of variables.

For the system shown in Figure 3.1, with the proposed input update law, we are interested in necessary and sufficient conditions for convergence of the iterative process. The following theorem provides a sufficient condition.

Theorem 3.1 *For the LTI plant $y_k = T_s u_k$, the proposed LTI learning control algorithm*

$$u_{k+1} = T_u u_k + T_e(y_d - y_k)$$

converges to a fixed point $u^*(t)$ if

$$\|T_u - T_e T_s\|_i < 1.$$

The fixed point u^* is given by

$$u^*(t) = (I - T_u + T_e T_s)^{-1} T_e y_d(t)$$

and the resulting fixed point of the error is given by

$$e^*(t) = \lim_{k \to \infty} (y_k - y_d) = (I - T_s (I - T_u + T_e T_s)^{-1} T_e) y_d(t),$$

for $t \in (t_0, t_f)$.

Proof :

First, notice that the next input u_{k+1} can be written as

$$u_{k+1} = T_u u_k + T_e (y_d - y_k) = T_u u_k + T_e (y_d - T_s u_k) = f(u_k)$$

and consequently,

$$\|f(x_1) - f(x_2)\| = \|T_u(x_1 - x_2) - T_e T_s(x_1 - x_2)\| \leq \|T_u - T_e T_s\|_i \|x_1 - x_2\|.$$

This implies that the induced operator norm $\|T_u - T_e T_s\|_i$ is a Lipschitz constant for the iteration. If this quantity has value less than one, then we can guarantee convergence to a fixed point $u^*(t)$. Next, to show the form of $u^*(t)$ and $e^*(t)$, suppose the Lipschitz condition is satisfied. Then in the limit u^* must satisfy

$$u^* = T_u u^* + T_e (y_d - T_s u^*).$$

Combining terms, we obtain

$$u^*(t) = (I - T_u + T_e T_s)^{-1} T_e y_d(t).$$

Note that the fact that $\|T_u - T_e T_s\|_i < 1$, coupled with the small-gain theorem, assures us that the inverse in this equation exists. Finally, to obtain the fixed point of the error, we substitute the expression for u^* into

$$e^*(t) = y_d(t) - y^*(t) = y_d(t) - T_s u^*(t)$$

to get

$$e^*(t) = (I - T_s (I - T_u + T_e T_s)^{-1} T_e) y_d(t).$$

25

QED

Now we have a condition for convergence and an expression for the fixed point of the error. There are two questions to ask about this expression. One, we want to know when, if ever, the final error $e^*(t)$ is identically zero over the interval. Second, if the error is never zero, then we would like to know how could we make it as small as possible. We will address these questions in the next two sections.

3.1 CONVERGENCE WITH ZERO ERROR

From the expression

$$e^*(t) = (I - T_s(I - T_u + T_eT_s)^{-1}T_e)y_d(t),$$

we can see that in general the error $e^*(t)$ is not zero over the entire interval of interest. However, if we choose $T_u = I$, then our learning controller algorithm becomes

$$u_{k+1} = u_k + T_e e_k.$$

If the learning control scheme converges then the corresponding fixed point must satisfy

$$u^* = u^* + T_e e^*.$$

This implies that the final error satisfies $e^*(t) = 0$, for all $t \in (t_0, t_f)$, or equivalently,

$$\lim_{k \to \infty} y_k = y_d.$$

From Theorem 3.1, if $T_u = I$, then a sufficient condition is

$$\|I - T_eT_s\|_i < 1.$$

This is similar to many convergence conditions contained in the literature. Unfortunately, this condition can be overly restrictive, sometimes leading to a requirement that the plant be invertible. One example of this difficulty is found when

26

we impose an L_2 optimality criterion on our learning control problem. That is, we wish to zero the energy of the error, defined by the L_2-norm

$$\|x(t)\|_{L_2} = \left(\int_0^\infty x(t)^T x(t) dt \right)^{1/2}.$$

Suppose that the signals defined in Figure 3.1 are defined as $u \in U = L_2^r(0, \infty)$ and $y \in Y = L_2^m(0, \infty)$ (i.e., $u \in U$ is r-dimensional and $y \in Y$ is m-dimensional). Because $L_2(0, \infty)$ is isomorphically isometric to the space H_2, we may work in the frequency domain, with operators that are transfer matrices in the space H_∞. (The space H_2, called a Hardy space, is the normed linear vector space of all functions of a complex variable that are analytic (the function and all its derivatives exist) in the right half of the complex plane. H_∞ is the space of all stable transfer functions). Thus our learning control algorithm can be written as

$$U_{k+1}(s) = U_k(s) + H_e(s)[Y_d(s) - Y_k(s)],$$

where $H_e(s)$ is a transfer matrix. Define the system by $H_s(s)$. Then Theorem 3.1 says that $y_k(t) \to y_d(t)$ in the sense of the L_2-norm if

$$\|I - H_e(s)H_s(s)\|_\infty < 1,$$

where the H_∞-norm of a transfer matrix $H(s)$ is defined as

$$\|H(s)\|_\infty = \sup_\omega \lambda_{max}^{\frac{1}{2}}\{H^*(j\omega)H(j\omega)\}.$$

In words, the H_∞-norm is the maximum singular value of $H(s)$ over the $j\omega$-axis.

This is a good way to formulate the problem, because the class $L_2(0, \infty)$ can be interpreted as the class of bounded energy signals and thus contains many signals of interest. Also, there exist results from H_∞ control theory for solving optimization problems involving the H_∞ norm [57,58]. Unfortunately, it turns out that the convergence condition in this topology is satisfied only if the plant $H_s(s)$ is invertible. This can be seen by taking a closer look at the basic definitions, which yields the following theorem.

Theorem 3.2 *Given $H_s(s)$, there exists a proper, stable, LTI solution, $H_e(s) \in H_\infty$, to the problem*

$$\|I - H_e(s)H_s(s)\|_\infty < 1,$$

if and only if $H_s(s)$ is invertible over the space H_∞.

27

Proof:

If $H_s^{-1}(s) \in H_\infty$ then simply let $H_e(s) = H_s^{-1}(s)$ so that

$$\|I - H_e(s)H_s(s)\|_\infty = 0 < 1,$$

which shows the sufficiency. To show necessity there are two cases to consider. If $H_s(s)$ is not invertible then either (i) $H_s(s)$ is strictly proper and/or (ii) $H_s(s)$ is not minimum phase. We consider each of these cases separately.

(i) If $H_s(s)$ is strictly proper, then $\lim_{\omega \to \infty} H_s(j\omega) = 0$, and for any proper, LTI system $H_e(s)$ it will be true that $\lim_{\omega \to \infty} H_e(j\omega)H_s(j\omega) = 0$. Thus

$$
\begin{aligned}
\|I - H_e(s)H_s(s)\|_\infty &= \sup_\omega \lambda_{max}^{\frac{1}{2}}\{(I - H_e(j\omega)H_s(j\omega))^*(I - H_e(j\omega)H_s(j\omega))\} \\
&\geq \lambda_{max}^{\frac{1}{2}}\{(I - H_e(\infty)H_s(\infty))^*(I - H_e(\infty)H_s(\infty))\} \\
&\geq \lambda_{max}^{\frac{1}{2}}\{I\} = 1.
\end{aligned}
$$

(ii) If $H_s(s)$ is non-minimum phase, then there is a point $s_1 \in C^+$ such that $H_s(s_1) = 0$, and for any stable system $H_e(s)$ we have $H_e(s_1)H_s(s_1) = 0$. Thus

$$
\begin{aligned}
\|I - H_e(s)H_s(s)\|_\infty &= \sup_\omega \lambda_{max}^{\frac{1}{2}}\{(I - H_e(j\omega)H_s(j\omega))^*(I - H_e(j\omega)H_s(j\omega))\} \\
&= \sup_{s \in C^+} \lambda_{max}^{\frac{1}{2}}\{(I - H_e(s)H_s(s))^*(I - H_e(s)H_s(s))\} \\
&\geq \lambda_{max}^{\frac{1}{2}}\{(I - H_e(s_1)H_s(s_1))^*(I - H_e(s_1)H_s(s_1))\} \\
&\geq \lambda_{max}^{\frac{1}{2}}\{I\} = 1.
\end{aligned}
$$

So, there is no stable, proper, LTI system $H_e(s)$ that makes $\|I - H_e(s)H_s(s)\|_\infty$ less than unity if $H_s(s)$ is not invertible over the space H_∞. **QED**

This result shows that for the proposed class of learning controllers we must have an invertible plant to obtain convergence of the error in the sense of the L_2 norm. Recall that in the literature survey of Chapter 2 we pointed out a less general version of this theorem given by Mita and Kato in [30]. Their result considered the norm $\|I - aH_s(s)\|_\infty$. This is a special case of our theorem where $H_e(s) = a$.

Note that because we are using a contraction mapping approach to the learning control problem, Theorem 3.1 gives a condition which is sufficient, but not necessary. Even though there may be no stable, proper, LTI system T_e that satisfies

28

$\|I - T_e T_s\|_i < 1$, this does not imply that our original learning control problem is not solvable. With the exception of Mita and Kato's scheme, essentially all the schemes for LTI systems found in the literature use a learning control update law where the current input is not weighted (i.e., $T_u = I$). This is true even for the time-varying and the high-gain feedback schemes. In some of these schemes convergence to zero error is shown without a requirement of system invertibility. This can be attained in several ways. One way is to introduce different norms and enlarged classes of admissible learning controllers (such as the use of derivative action by Arimoto et al., resulting in a non-causal learning controller, or the time-varying controllers used by some of the researchers). Another observation is that for the final error to be zero we require $y_d = T_s u^*$. Thus y_d need only be in the image of T_s. This is a weaker condition than requiring that T_s be invertible. However, if T_s is not invertible, then it will be harder to ensure convergence of the learning iterations. In this case, a guarantee of convergence may not be possible, except in the context of a specific norm. Finally, in Chapter 5 we will see that when the final time t_f is finite some obvious simplifications can occur. However, the fact remains that for an L_2-optimality criterion, convergence with zero error for the proposed class of causal, LTI learning controllers is equivalent to a requirement of plant invertibility.

Another issue we have not addressed at this point concerns the plant knowledge required to design the learning controller T_e. Continuing the same example as above, for the L_2-optimality criterion we would need to know the model of the plant to solve the H_∞ problem for the learning controller transfer function $H_e(s)$. In fact, nearly all the schemes reported in the literature require some knowledge of the plant dynamics. We show in Chapter 4 that this difficulty can be avoided (for LTI systems) with the use of a learning control scheme based on parameter estimation.

3.2 CONVERGENCE WITH NON-ZERO ERROR

We now consider the more general form of the learning control algorithm with $T_u \neq I$. In this case, it is possible to avoid the problem of plant invertibility, while still guaranteeing convergence of the learning iterations. Specifically, let

$$u_{k+1} = T_u u_k + T_e(y_d - y_k).$$

Now the condition for convergence is $\|T_u - T_e T_s\|_i < 1$. This is much less restrictive than $\|I - T_e T_s\|_i < 1$, because we now have the freedom to pick T_u, in addition to T_e. However, now the error $e_k = (y_d - y_k)$ will not converge to zero. The task in this approach is to choose T_u and T_e so that the norm of the final error

$$e^*(t) = \lim_{k \to \infty} (y_d - y_k) = (I - T_s(I - T_u + T_e T_s)^{-1}T_e)y_d(t)$$

is minimized, while maintaining the convergence condition.

In this section we will continue to assume that we have full knowledge of the plant dynamics T_u. This will be necessary to minimize the norm of the final error, $\|e^*(t)\|$. Without some knowledge of the plant, our learning control scheme will be suboptimal. Our goal at this point is to examine the nature of the best possible solution to the learning control problem. Hence we suppose the plant is known. This gives us a "best case" standard that we can use as a benchmark for comparison when we relax the assumption.

For the following development, let U and Y be normed linear spaces and let X be the space of causal LTI operators mapping elements from U into Y. Our approach will be to cast our learning controller design problem as an optimization problem over the space X. We will then show that the solution of the optimization problem can be related to the solution of a type of model-matching problem. First consider the following optimization problem.

OPT1 : Let $u_k \in U$, $y_d, y_k \in Y$, and $T_s, T_u, T_e \in X$. Then given y_d and T_s, find T_u^* and T_e^* to solve the problem

$$\min_{T_u, T_e \in X} \|(I - T_s(I - T_u + T_e T_s)^{-1}T_e)y_d\|,$$

subject to

$$\|T_u - T_e T_s\|_i < 1.$$

The problem OPT1 seeks the learning control algorithm that will converge to a fixed point $e^*(t)$, with a norm that is the smallest that can possibly be achieved by any learning controller in the proposed class. In order to give the solution to OPT1 we must first introduce a second optimization problem, which is similar to a standard model-matching problem.

OPT2 : Let $y_d \in Y$ and let $T_n, T_s \in X$. Then given y_d and T_s, find T_n^* to solve the problem

$$\min_{T_n \in X} \|(I - T_s T_n) y_d\|.$$

OPT2 is the problem of finding the closest possible inverse of a system, in the direction of the desired input, with respect to a specified norm. The following theorem relates the solution of the learning control problem OPT1 to the solution of the inverse problem OPT2.

Theorem 3.3 *Let T_n^* be the solution of OPT2. Define T_e^* by the factorization $T_n^* = T_m^* T_e^*$, with T_m^* invertible over the space X and $\|I - T_m^{*-1}\|_i < 1$. Let $T_u^* = I - T_m^{*-1} + T_e^* T_s$. Then T_u^* and T_e^* are the solutions of OPT1.*

Proof:
If $T_u^* = I - T_m^{*-1} + T_e^* T_s$ and $T_e^* = T_m^{*-1} T_n^*$, then

$$\|(I - T_s(I - T_u^* + T_e^* T_s)^{-1} T_e^*) y_d\| = \|(I - T_s T_n^*) y_d\| = \min_{T_n \in X} \|(I - T_s T_n) y_d\|.$$

In addition, we may admit T_u^* and T_e^* as candidate solutions for the problem OPT1, because

$$\|T_u^* - T_e^* T_s\|_i = \|I - T_m^{*-1}\|_i < 1.$$

Now suppose $\bar{T}_u \neq T_u^*$ and $\bar{T}_e \neq T_e^*$ are solutions to OPT1, but that they do not admit the relationship defined by $\bar{T}_u = I - T_m^{*-1} + \bar{T}_e T_s$, with $T_n^* = T_m^* \bar{T}_e$ for some invertible operator T_m^*. Define $\bar{T}_m = (I - \bar{T}_u + \bar{T}_e T_s)^{-1}$. Note that $\bar{T}_m$ exists and is invertible because if $\bar{T}_u$ and $\bar{T}_e$ solve OPT1, then $\|\bar{T}_u - \bar{T}_e T_s\|_i < 1$. Also note that as T_u and T_e range over X, subject to $\|T_u - T_e T_s\|_i < 1$, then $T_m = (I - T_u + T_e T_s)^{-1}$ ranges over all invertible operators in X. Thus we may write

$$
\begin{aligned}
\min_{T_e, T_u \in X} \|(I - T_s(I - T_u + T_e T_s)^{-1} T_e) y_d\| &= \min_{T_e, T_m, T_m^{-1} \in X} \|(I - T_s T_m T_e) y_d\| \\
&= \|(I - T_s(I - \bar{T}_u + \bar{T}_e T_s)^{-1} \bar{T}_e) y_d\| \\
&= \|(I - T_s \bar{T}_m \bar{T}_e) y_d\|.
\end{aligned}
$$

31

Further, let $\bar{T}_n = \bar{T}_m\bar{T}_e$ and note that as T_e and T_m range over X, with T_m invertible, then $T_n = T_mT_e$ ranges over X. So we may write

$$\min_{T_e,T_m,T_m^{-1}\in X} \|(I - T_sT_mT_e)y_d\| = \min_{T_n\in X} \|(I - T_sT_n)y_d\|$$
$$= \|(I - T_s\bar{T}_m\bar{T}_e)y_d\|$$
$$= \|(I - T_s\bar{T}_n)y_d\|.$$

But, this implies that

$$\|(I - T_s\bar{T}_n)y_d\| < \|(I - T_sT_n^*)y_d\|,$$

which is a contradiction (equality is not a possibility because of the assumption that $\bar{T}_e$ and $\bar{T}_u$ are not related by the factorizations given in the theorem). Hence the theorem follows. **QED**

There are several points worth noting about this theorem. First, what Theorem 3.3 says is that, to solve the causal, LTI learning control problem OPT1, we simply solve the problem OPT2. This is just the problem of finding an approximate inverse in a specified topology. This is a useful result because it is often possible to solve these types of problems. For instance, if we choose an l_∞-optimality criterion (to minimize the maximum value of the error), then OPT2 can be solved by some recent results of Dahleh and Pearson [3]. Similarly, given an L_2-optimality criterion (to minimize the energy of the error), we can use well-established methods from L_2 control theory to find T_e^* and T_u^*.

Second, the factorization in Theorem 3.3 is not unique, as we will discuss below. Hence we may exhibit more than one solution to OPT1. However, the operators T_e^* and T_u^* must be related in such a way that the error attains its minimum norm. That is, they must be related by the definitions given in the theorem. This can be seen from the proof. If they are not, then the error does not attain its minimum norm. This is why it will be difficult to make the error converge to its smallest possible value if we have no knowledge of the plant dynamics. Without such information, it is unlikely that we would pick the operators T_e and T_u properly, and then we would end up with a sub-optimal learning controller.

A third note is that we have kept the desired output y_d fixed in our problem formulation. An obvious extension would be to consider the problem when we wish to minimize the final error $e^*(t)$ for any input. In this case our problem becomes the following:

Given T_s, find T_u and T_e to solve the problem

$$\min_{T_u, T_e \in X} \|I - T_s(I - T_u + T_e T_s)^{-1} T_e\|_i,$$

subject to

$$\|T_u - T_e T_s\|_i < 1.$$

We can show that this is equivalent to the problem

$$\min_{T_n \in X} \|I - T_s T_n\|_i.$$

This is in the form of a standard model-matching problem and can be solved using results from H_∞ or L_1 control theory [57,59], depending on the norms that are chosen. Such an approach would correspond to a worst case design philosophy for the learning control problem.

3.3   THE NATURE OF THE SOLUTION

The implications of Theorem 3.3 give us some insight into the nature of the solution of the learning control problem. Assume that we have obtained T_n^*, the solution to problem OPT2. To find T_u^* and T_e^* we must first factor $T_n^* = T_m^* T_e^*$, with T_m^* invertible and $\|I - T_m^{*-1}\| < 1$. This can always be done, because we can choose $T_m^* = I$. This leads to two possible cases: (i) $T_m^* = I$ and (ii) $T_m^* \neq I$.

(i) If we choose $T_m^* = I$, then Theorem 3.3 tells us that $T_e^* = T_n^*$ and $T_u^* = T_e^* T_s$. Thus the optimal input update algorithm becomes

$$\begin{aligned}
u_{k+1} &= T_u^* u_k + T_e^* (y_d - y_k) \\
&= T_u^* u_k + T_e^* (y_d - T_s u_k) \\
&= T_e^* T_s u_k + T_n^* y_d - T_e^* T_s u_k \\
&= T_n^* y_d(t).
\end{aligned}$$

But this implies that

$$u^*(t) = u_{k+1} = u_k = T_n^* y_d(t),$$

33

which shows that our input does not change from trial to trial. Hence, no iterative learning is required and the minimum possible norm of the error is attained by using an input defined by $u^*(t) = T_n^* y_d(t)$. This can also be seen by substituting $T_e^* = T_n^*$ and $T_u^* = T_n^* T_s$ into

$$e^*(t) = (I - T_s(I - T_u + T_e T_s)^{-1} T_e) y_d(t),$$

to obtain

$$e^*(t) = (I - T_s T_n^*) y_d(t).$$

But, by the definition of T_n^*, this error will have minimum norm. Notice that the same expression for u^* can also be obtained by substituting $T_e^* = T_n^*$ and $T_u^* = T_n^* T_s$ into the expression from Theorem 3.1, to get

$$
\begin{aligned}
u^*(t) &= (I - T_u + T_e T_s)^{-1} T_e y_d(t) \\
&= (I - T_n^* T_s + T_n^* T_s)^{-1} T_n^* y_d(t) \\
&= T_n^* y_d(t).
\end{aligned}
$$

(ii) Next suppose that when we factor $T_n^* = T_m^* T_e^*$, we choose $T_m^* \neq I$, but satisfying T_m^* invertible and $\|I - T_m^{*-1}\|_i < 1$. Let $T_u^* = I - T_m^{*-1} + T_e^* T_s$ as defined in Theorem 3.3. Then applying T_u^* and T_e^* to the expressions for the fixed point u^* and the resultant error e^* from Theorem 3.1, we obtain

$$
\begin{aligned}
u^*(t) &= (I - T_u^* + T_e^* T_s)^{-1} T_e^* y_d(t) \\
&= (I - (I - T_m^{*-1} + T_e^* T_s) + T_e^* T_s)^{-1} T_e^* y_d(t) \\
&= T_n^* y_d(t)
\end{aligned}
$$

and

$$e^*(t) = y_d(t) - T_s u^*(t) = (I - T_s T_n^*) y_d(t).$$

These are exactly the same expressions we obtained for the fixed points of the input and error signals when $T_m^* = I$. That is, the optimal input $u^*(t)$, and the associated error $e^*(t)$, are the same for both cases. This implies that the best input signal for the system, when the plant is known, can be determined either through a learning process or through an *a priori* design (by best we mean best in terms of minimizing the norm of the final error signal). In fact, to obtain the best possible learning control law, it is first necessary to solve the *a priori* design problem to obtain T_n^*, from which T_u^* and T_e^* can be derived. Thus, we conclude that for the class of LTI learning controllers, learning is not really necessary if the plant is known. Furthermore, any learning controller in this class will be suboptimal if it is designed without knowledge of the plant parameters. These comments are summarized below for emphasis.

Conclusion: For a causal, LTI, stable system S described by $y = T_s u$, with T_s known, the best possible learning controller, from the class of causal, LTI learning controllers defined by

$$u_{k+1} = T_u u_k + T_e(y_d - y_k),$$

can do no better than the open-loop control law

$$u^*(t) = T_n^* y_d(t),$$

where T_n^* is the best approximation to the inverse system of T_s for the input $y_d(t)$ and is defined as the solution of the problem

$$\min_{T_n \in X} \|(I - T_s T_n) y_d)\|.$$

CHAPTER 4

LTI LEARNING CONTROL VIA PARAMETER ESTIMATION

In the previous chapter we were interested in a qualitative assessment of the nature of a linear time-invariant (LTI) learning scheme. Because we wanted to get some idea of the best possible performance, we assumed the system dynamics were known. In this chapter we partially relax this assumption. We now consider the learning control problem when the plant has a known structure, but unknown parameters. Our approach to this problem will be based on parameter estimation techniques. Previously we found that, if the plant is known, then iteration is not necessary. However, if the plant is not known, then it seems reasonable to expect that learning would be useful in improving the system response. Below we describe a learning control scheme for this situation. We begin with a description of the scheme. Then we present our main result, followed by some comments on the implications of the result.

4.1 SYSTEM DESCRIPTION

As in the previous chapter, let the plant be a causal, LTI, stable system. We also make the following assumptions: (i) the order of the plant is known, (ii) the plant is minimal (i.e., controllable and observable), (iii) the plant is single-input, single-output, and (iv) the plant is a discrete-time system. Note, however, that the development also follows for multiple-input, multiple-output systems, as well as for continuous-time systems.

For a plant satisfying these assumptions, a proposed learning control scheme is shown in Figure 4.1. The learning controller operates in the following way. First a parameter estimator is used to obtain estimates of the system during each trial. These estimates are then used to solve the problem OPT2 as described in the previous chapter. Finally, the solution of OPT2 is used as the optimal input for the next trial. This procedure is repeated on each learning iteration.

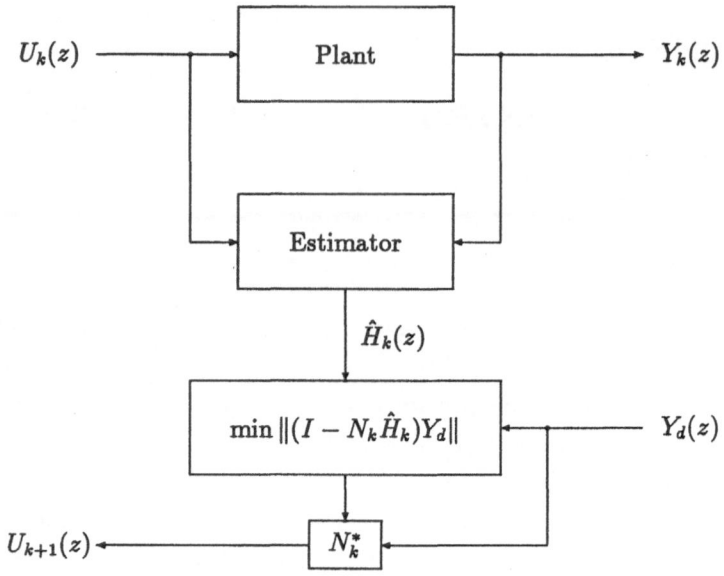

Figure 4.1: Learning control using a parameter estimator.

In this chapter we discuss the convergence properties of this scheme for the case of a least-squares estimator, using an l_∞-optimality criterion in the problem OPT2. However, neither of these constraints are essential, and the results would hold in general for any norm and for any estimation scheme that converges. In this section we first give some notation. Then we describe the parameter estimator and the learning control law. We give our main result in the next section.

4.1.1 Notation

Let $X(z)$ denote the z-transform of the time signal $x = x(i)$, where i is an integer variable representing time and $x \in l_\infty$, the class of bounded sequences. The z-transform is defined in the usual way by

$$X(z) = \sum_{i=0}^{\infty} x(i)z^{-i}.$$

Let A be the class of stable, rational functions in z and let the system be defined by the transfer function $H(z) \in A$, where

$$H(z) = \frac{b_1 z^{n-1} + \cdots + b_{n-1} z + b_n}{z^n + a_1 z^{n-1} + \cdots + a_{n-1} z + a_n}.$$

Again we comment that, if our original system is not stable, then we must first stabilize it with a conventional controller before applying the learning control scheme.

We will now parameterize our plant in the time-domain, using standard notation from the literature of system identification. Specifically, suppose that at the kth trial we have $Y_k(z) = H(z)U_k(z)$. Expressing this in the time-domain, we can write the difference equation

$$y_k(i) = -\sum_{j=1}^{n} a_j y_k(i-j) + \sum_{j=1}^{n} b_j u_k(i-j).$$

If we introduce a vector made up of the system parameters,

$$\theta_0 = (-a_1, -a_2, \cdots, -a_n, b_1, b_2, \cdots, b_n)^T,$$

and we introduce the following sequence of regression vectors

$$\phi_k(i-1) = (y_k(i-1), y_k(i-2), \cdots, y_k(i-n), u_k(i-1), u_k(i-2), \cdots, u_k(i-n))^T,$$

then we can write the system output signal $y_k(i)$ in the time-domain as

$$y_k(i) = \theta_0^T \phi_k(i-1) = \phi_k^T(i-1)\theta_0.$$

The idea of our scheme is that the estimator in Figure 4.1 will produce estimates $\hat{\theta}_k(i)$ of the parameter vector θ_0, which can then be used to obtain an estimate of the plant transfer function.

4.1.2 Parameter Estimator and Learning Control Law

The learning control scheme we propose is motivated by the results from Chapter 3. The input update algorithm is defined by

$$U_{k+1}(z) = N_k^*(z)Y_d(z),$$

where $N_k^*(z)$ is the solution to the l_∞ minimization problem

$$\min_{N_k \in A} \|(I - N_k(z)\hat{H}_k(z))Y_d(z)\|_{l_\infty}.$$

In this minimization problem, $\hat{H}_k(z)$ is the final estimate of the transfer function of the plant at the kth trial. This estimate is constructed from the elements of

$$\hat{\theta}_k(N) = (-\hat{a}_1, -\hat{a}_2, \cdots, -\hat{a}_n, \hat{b}_1, \hat{b}_2, \cdots, \hat{b}_n)^T$$

as

$$\hat{H}_k(z) = \frac{\hat{b}_1 z^{n-1} + \cdots + \hat{b}_{n-1} z + \hat{b}_n}{z^n + \hat{a}_1 z^{n-1} + \cdots + \hat{a}_{n-1} z + \hat{a}_n}.$$

In other words, we take $\hat{\theta}_k(N)$, the final estimate of the plant parameters at the kth trial, and use this to construct an estimate of the plant transfer function. This is then used in the l_∞ problem to obtain $N_k^*(z)$, which is used to form the input for the next trial. (Actually, by the results of [3], the l_∞ problem cannot be solved exactly, but it is possible to obtain a sub-optimal solution, which can be made arbitrarily close to the optimal solution.)

As noted above, for this presentation we will assume a least-square estimator (possibly recursive). However, the development is really generic and similar results can be expected to hold for other estimators. The following rule is used to estimate the parameter vector during each trial k [60]. For $i = 1, 2, \cdots, N$, let

$$\hat{\theta}_k(i) = \hat{\theta}_k(i-1) + \frac{P_k(i-2)\phi_k(i-1)}{1 + \phi_k^T(i-1)P_k(i-2)\phi_k(i-1)}[y_k(i) - \phi_k^T(i-1)\hat{\theta}_k(i-1)],$$

$$P_k(i-1) = P_k(i-2) - \frac{P_k(i-2)\phi_k(i-1)\phi_k^T(i-1)P_k(i-2)}{1 + \phi_k^T(i-1)P_k(i-2)\phi_k(i-1)},$$

$$P_{k+1}(-1) = P_k(N-1),$$

$$\hat{\theta}_{k+1}(0) = \hat{\theta}_k(N),$$

where $P_0(-1)$ is an arbitrary positive definite matrix, $\hat{\theta}_1(0)$ is a given initial estimate of the parameter vector, and each trial is conducted over the finite interval $[0, N]$.

The key to this learning control scheme is the manner in which the estimator is initialized at the start of each trial. The important idea is that at the start of each trial we initialize the estimator with the final covariance update from the previous trial. That is, the parameter vector that was obtained at the end of a given trial is then used to initialize the estimation algorithm at the beginning of the next trial. This is the property that makes convergence of the learning control iterations possible.

4.2   MAIN RESULT

Before we give the main result of this chapter we state two technical results that are necessary for the proof of the result. The first concerns the issue of persistence of excitation [60]. Persistence of excitation means that the input to our system is "rich" enough to excite all the important dynamical modes of the system. In any control algorithm based on system identification techniques, our input signal must be persistently exciting to ensure that our parameter estimates will converge to the correct values. The second technical result is concerned with the continuity of the solutions of an l_∞-minimization problem.

Lemma 4.1 *For the learning control scheme described above, the sequence of inputs $U_k(z)$ is persistently exciting of order $2n$ if the initial input $U_0(z)$ is persistently exciting of order $2n$.*

Lemma 4.2 *If $\lim_{k\to\infty} \hat{H}_k(z) = H(z)$, then $\lim_{k\to\infty} N_k^*(z) = N^*(z)$, where $N_k^*(z)$ is the solution of the l_∞ minimization problem*

$$\min_{N_k(z)\in A} \|(I - N_k(z)\hat{H}_k(z))Y_d(z)\|_{l_\infty}$$

and $N^*(z)$ is the solution to the problem

$$\min_{N(z)\in A} \|(I - N(z)H(z))Y_d(z)\|_{l_\infty}.$$

We may now state the main result of this chapter.

Theorem 4.1 *For the configuration of Figure 4.1, with the plant, estimator, and learning control law as described above, let the initial input u_0 be persistently exciting of order $2n$ or more. Then*

$$\lim_{k\to\infty} \hat{H}_k(z) = H(z),$$

41

which implies that

$$U^*(z) = \lim_{k \to \infty} N_k^*(z) Y_d(z) = N^*(z) Y_d(z),$$

where $N^*(z)$ is the solution of the inverse problem

$$\min_{N(z) \in A} \|(I - N(z) H(z)) Y_d(z)\|_{l_\infty}.$$

This implies that the sequence of input signals $U_k(z)$ converges to the best possible input to the system in terms of minimizing the l_∞-norm of the error $e^*(i) = y_d(i) - y^*(i)$.

Proof:

Note that if we show

$$\lim_{k \to \infty} \hat{H}_k(z) = H(z),$$

then given Lemma 4.2 and Theorem 3.3 the result follows. To show

$$\lim_{k \to \infty} \hat{H}_k(z) = H(z),$$

observe that each trial has a finite duration of N samples. Thus, consider a sequence of vectors defined by

$$\hat{\theta}(i) = \{\hat{\theta}_1(0), \cdots, \hat{\theta}_1(N), \hat{\theta}_2(0), \cdots, \hat{\theta}_2(N), \cdots, \hat{\theta}_k(0), \cdots, \hat{\theta}_k(N), \cdots\}.$$

That is,

$$\hat{\theta} = \begin{cases} \hat{\theta}_k(i - (k-1)N) & \text{if } (k-1)N < i \leq kN \\ \hat{\theta}_1(0) & \text{if } i = 0. \end{cases}$$

Similarly, for $1 \leq (k-1)N + 1 \leq i \leq kN$, define

$$\phi(i - 1) = \phi_k(i - (k-1)N)$$

and

$$P(i - 1) = \begin{cases} P_k(i - (k-1)N - 1) & \text{if } (k-1)N + 1 \leq i \leq kN \\ P_1(-1) & \text{if } i = 0. \end{cases}$$

42

Now simply note that the sequence $\hat{\theta}(i)$ can be viewed as the output of a least squares estimator, with a sequence of regression vectors $\phi(i-1)$ and associated covariance matrices $P(i-1)$. This is a result of the way we have defined the initial conditions $\hat{\theta}_k(0)$ and $P_k(-1)$ at each trial. Then, by properties of the least-squares algorithm and our assumptions of stability, minimality, and persistence of excitation of the original input (which, by Lemma 4.1, ensures persistence of excitation of the input sequence), we have [60]

$$\lim_{i \to \infty} \hat{\theta}(i) = \theta_0.$$

But, if $\hat{\theta}(i)$ converges then so does any subsequence. In particular, consider the subsequence defined by

$$\hat{\theta}(i_k) = \hat{\theta}(kN) = \hat{\theta}_k(N),$$

for $k = 1, 2, \cdots$. Then

$$\lim_{k \to \infty} \hat{\theta}_k(N) = \theta_0,$$

and thus

$$\lim_{k \to \infty} \hat{H}_k(z) = H(z).$$

QED

4.3   COMMENTS

Theorem 4.1 tells us that the proposed learning control scheme based on parameter estimates will converge to an error signal that has a minimum l_∞-norm. Thus it is not necessary to have full knowledge of the plant to design the best possible LTI learning controller for an l_∞-optimality criterion (or any other optimality criterion, for that matter). However, a similar comment can be made about the capability of a conventional adaptive control scheme.

This claim can be understood by referring back to Figure 4.1. Suppose we execute only a single trial, but allow the trial to last for a very long time. If we keep the same assumptions on our plant and our control scheme (e.g., persistence of excitation during the trial), then we can still use the least-squares estimator

43

to obtain an estimate of the parameter vector that converges asymptotically to the true value. In this case the error in the estimate at time $i_k = kN$ will be the same estimation error we have at the end of the kth-trial in the learning control scheme. Hence, instead of k trials of duration N, we only need to execute one trial of duration kN. From this trial, we will be able to obtain a convergent estimate of the plant, which can then be used to solve the l_∞ problem, resulting in the best possible input as described in Theorem 3.3 (again, all these comments apply for any optimality criterion). So, as in the case when we have full plant knowledge, we conclude that no iterative learning is required. This is summarized below for emphasis.

Conclusion: Consider a discrete-time, causal, LTI, minimal system S, described by $Y(z) = H(z)U(z)$, where the order of $H(z)$ is known, but otherwise $H(z)$ is unknown. For this system, the learning control scheme depicted in Figure 4.1 converges to the open-loop control law

$$U^*(z) = N^*(z)Y_d(z),$$

where $N^*(z)$ is the best approximation to the inverse system of $H(z)$ for the input $Y_d(z)$ and is defined as the solution to the l_∞-minimization problem

$$\min_{N \in A} \|(I - NH)Y_d\|_{l_\infty}.$$

However, the same control law can be obtained by executing a single trial of a long duration and using a conventional parameter estimator. At the end of the single trial the parameter estimate can be used to solve the l_∞ problem to obtain the same open-loop control law given by the iterative learning control scheme. Hence the approach of iterative learning control offers no advantage over existing techniques for this class of systems.

CHAPTER 5

FINITE-HORIZON LEARNING CONTROL

In the two preceding chapters we have presented a general development of the learning control problem for linear time-invariant (LTI) systems. The results presented can be applied to both finite- and infinite-horizon problems. However, as we have noted, any practical implementation of an iterative learning controller will be a finite-time problem. That is, the duration of each trial will be fixed to some time less than infinity. This is inherent in the nature of repetitive operations. In addition, our formulation and results are equally valid for continuous-time and discrete-time systems. However, to implement a learning control scheme we will have to use a microprocessor-based controller. For these reasons, it is reasonable to restrict our attention to discrete-time plants that are operated repetitively on a finite time horizon. In this chapter we consider the learning control problem for such systems. Our results are given for LTI, single-input, single-output plants, but can be generalized to multiple-input, multiple-output systems. We first give a learning control scheme with memory, using an l_∞ criterion. Then we show that this scheme can actually be modified to give a single-trial convergence rate. In the third section we present a learning control scheme for a discrete-time system with multirate sampling. These techniques are illustrated with examples. The final section of the chapter discusses extensions of these results to linear time-varying systems.

5.1 l_∞-OPTIMAL LEARNING CONTROL WITH MEMORY

Consider Figure 5.1, which presents a slightly different block diagram description of a learning control system (cf. Figure 3.1). Let the LTI system have relative degree m (i.e., it has m more poles than zeros), so that we can describe its transfer function (using z-transforms) by

$$P(z) = z^{-m} H(z).$$

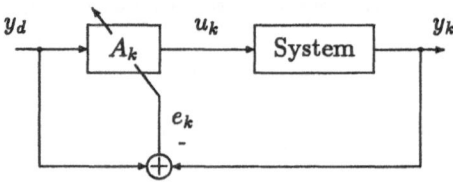

Figure 5.1: Inverse system estimator.

For example,

$$P(z) = \frac{z+1}{z^3 + .707z^2 + .707z + 1} = z^{-2}\frac{z^3 + z^2}{z^3 + .707z^2 + .707z + 1}$$

It is easy to show that m is the number of time steps between the first non-zero input to the system and the first non-zero output. Also notice that $H(z)$ has the same number of poles as zeros.

The input to the plant at each trial is computed according to $u_k = A_k y_d$, where the system A_k is an estimate of the approximate inverse of the plant at the kth trial. This system will be updated (as described below) after each trial and then used to compute the input for the next trial. Hence we can think of A_k as the learning controller. To update the system A_k, we notice that the first m outputs of the plant can never be changed by choice of the input. Thus our update algorithm will use only the errors on the interval $[m, m + N]$, where N is the length of the desired trajectory. The effect of our algorithm is to invert $H(z)$ on this interval, using input/output measurements.

From the trajectories of the input, output, and desired output, define the vectors

$$
\begin{aligned}
u_k &= (u_k(0), u_k(1), \cdots, u_k(N-1)), \\
y_k &= (y_k(m), y_k(m+1), \cdots, y_k(N-1+m)), \\
y_d &= (y_d(m), y_d(m+1), \cdots, y_d(N-1+m)).
\end{aligned}
$$

The subscript denotes the kth trial. With no loss of generality, we will set $m = 1$. Also, notice that the signals u_k, y_k, y_d, and the error $e_k = y_d - y_k$ are all in the space R^N. Thus for our analysis we can use the truncated l_∞-norm, given by

$$\|x\|_\infty = \max_{1 \le i \le N} |x_i|.$$

46

This norm measures the value of the maximum magnitude component of $x \in R^N$. The corresponding induced norm for a linear operator $H : R^N \rightarrow R^N$ is given by

$$\|H\|_i = \|H\|_\infty = \max_i (\sum_{j=1}^{N} |h_{ij}|).$$

This norm computes the maximum row sum of the matrix.

Using this notation, our linear plant can be described by $y_k = H u_k$, where H is a matrix of rank N whose elements are the Markov parameters of $H(z)$ (here we use the symbol H to denote a matrix associated with the transfer function $H(z)$). H is written as the lower triangular matrix

$$H = \begin{bmatrix} h_1 & 0 & 0 & \cdots & 0 \\ h_2 & h_1 & 0 & \cdots & 0 \\ h_3 & h_2 & h_1 & \cdots & 0 \\ \vdots & \vdots & \vdots & \ddots & \vdots \\ h_N & h_{N-1} & h_{N-2} & \cdots & h_1 \end{bmatrix}.$$

Note that the matrix H is Toeplitz. That is, it is constant along its diagonals. This makes it convenient to perform computations using H. Also, the Markov parameters which define the entries of H can be computed by expanding the transfer function $H(z)$ in powers of z^{-1} as

$$H(z) = \sum_{k=1}^{\infty} h_k z^{-k},$$

where, by previous assumptions on $H(z)$ we have $h_0 = 0$.

Now, the input to the plant is defined by

$$u_k = A_k y_d,$$

where the approximate inverse matrix A_k is also a Toeplitz matrix. Let A_k be updated at the end of each trial according to

$$A_{k+1} = A_k + \Delta A_k,$$

with

$$\Delta A = \begin{bmatrix} \alpha_1 e_k(1) & 0 & 0 & \cdots & 0 \\ \alpha_1 e_k(2) & \alpha_2 e_k(1) & 0 & \cdots & 0 \\ \alpha_1 e_k(3) & \alpha_2 e_k(2) & \alpha_3 e_k(1) & \cdots & 0 \\ \vdots & \vdots & \vdots & \ddots & \vdots \\ \alpha_1 e_k(N) & \alpha_2 e_k(N-1) & \alpha_3 e_k(N-2) & \cdots & \alpha_N e_k(1) \end{bmatrix}.$$

47

That is,

$$\Delta a_{ij} = \alpha_j e_k(i - j + 1),$$

for $i = 1, \cdots, N$ and $j = 1, \cdots, i$. The α_i's are design parameters that govern the convergence properties of the algorithm.

This learning control algorithm can be rewritten in the standard form used in Chapter 3 as follows. From Figure 5.1, we have

$$
\begin{aligned}
u_{k+1} &= A_{k+1} y_d = (A_k + \Delta A_k) y_d \\
&= u_k + \Delta A_k y_d.
\end{aligned}
$$

But,

$$
\Delta A y_d =
\begin{bmatrix}
\alpha_1 e_k(1) & 0 & 0 & \cdots & 0 \\
\alpha_1 e_k(2) & \alpha_2 e_k(1) & 0 & \cdots & 0 \\
\alpha_1 e_k(3) & \alpha_2 e_k(2) & \alpha_3 e_k(1) & \cdots & 0 \\
\vdots & \vdots & \vdots & \ddots & \vdots \\
\alpha_1 e_k(N) & \alpha_2 e_k(N-1) & \alpha_3 e_k(N-2) & \cdots & \alpha_N e_k(1)
\end{bmatrix}
y_d,
$$

which can be rewritten as

$$
\Delta A y_d =
\begin{bmatrix}
\alpha_1 y_d(1) & 0 & \cdots & 0 \\
\alpha_2 y_d(2) & \alpha_1 y_d(1) & \cdots & 0 \\
\alpha_3 y_d(3) & \alpha_2 y_d(2) & \cdots & 0 \\
\vdots & \vdots & \ddots & \vdots \\
\alpha_N y_d(N) & \alpha_{N-1} y_d(N-1) & \cdots & \alpha_1 y_d(1)
\end{bmatrix}
e_k = H_e e_k.
$$

Thus, our learning law has the form

$$u_{k+1} = u_k + H_e e_k.$$

This is now the standard form that we used for LTI learning control in the previous chapters. Hence, by Theorem 3.1 this will converge with zero error whenever $\|I - H_e H\|_\infty < 1$. But this can always be achieved by choice of the α_i's because H is Toeplitz with $h_1 \neq 0$, which implies that H is invertible. Specifically, let

$$
A = H^{-1} =
\begin{bmatrix}
a_1 & 0 & 0 & \cdots & 0 \\
a_2 & a_1 & 0 & \cdots & 0 \\
a_3 & a_2 & a_1 & \cdots & 0 \\
\vdots & \vdots & \vdots & \ddots & \vdots \\
a_N & a_{N-1} & a_{N-2} & \cdots & a_1
\end{bmatrix}.
$$

Note that we can write the a_i's in terms of the h_i's as

$$a_i = \frac{1}{h_1} \sum_{j=1}^{i-1} a_j h_{i-j+1}.$$

Therefore we simply pick the α_i's so that $H_e = H^{-1}$. This is obviously satisfied by the choice $\alpha_i = a_i/y_d(i)$.

With this choice of the α_i's we can ensure that

$$\lim_{k \to \infty} \|e_k\|_\infty = 0.$$

Further, because this is pointwise convergence, this implies

$$\lim_{k \to \infty} y_k = \lim_{k \to \infty} H u_k = \lim_{k \to \infty} H A_k y_d = y_d.$$

This holds for any y_d, so we conclude that $\lim_{k \to \infty} A_k = H^{-1}$. Thus the sequence of matrices A_k converges to the inverse of the plant on the interval. We have just proven the following theorem.

Theorem 5.1 *For the configuration of Figure 5.1, with the learning control law given above, there exist gains α_i, $i = 1, \cdots, N$ such that*

$$\lim_{k \to \infty} \|e_k\|_\infty = 0,$$

which implies that

$$\lim_{k \to \infty} A_k = H^{-1}.$$

This algorithm provides an approach for convergence in the sense of the l_∞ norm. It also results in a memory feature. Specifically, after convergence we have available the matrix $A = H^{-1}$. If we then want to compute the best input signal for any other desired trajectory, we need only pass the desired trajectory through the matrix A. Because there is no need to repeat the iterative learning process, we call this scheme "learning with memory." Other learning schemes reported in the literature do not demonstrate this ability.

Note that Arimoto et al.'s original scheme in the discrete-time is a special case of what we have presented here. In their original result, they required $CB \neq 0$, where C and B were the output and input coupling matrices of the state space description, respectively. This is equivalent to the case $m = 1$ (i.e., the transfer function has a pole-zero excess of one, which gives a one-step delay through the system). In this case the discrete-time version of their result uses the learning law

$$u_{k+1} = u_k + \Gamma e_k(i+1).$$

But this is a special case of our result, with $\Gamma = \alpha_1 I$ and $\alpha_2 = \alpha_3 = \cdots = \alpha_N = 0$.

Our result also tells us how to proceed when $CB = 0$. In this case, as we suggested in Chapter 2, we find the smallest integer j such that $CA^{j-1}B \neq 0$. This is equivalent to having $m = j$. Thus we could use a learning law of the form

$$u_{k+1} = u_k + \Gamma e_k(i+m).$$

That is, we are effectively using the higher derivatives to account for the delay through the plant. Of course, this learning law does not converge "with memory" (although, analogous to Theorem 5.1, we can devise a scheme which does converge with memory for this case). For this reason, such learning control schemes have been described as perturbation learning methods [56]. The continuous-time version of the same result would be

$$u_{k+1}(t) = u_k(t) + \Gamma \frac{d^j}{dt^j} e_k(t).$$

Hence we have generalized Arimoto et al.'s original result.

5.2   LEARNING CONVERGENCE IN ONE STEP

In order to ensure convergence using the learning control scheme of the previous section, one must choose the correct gains α_i, so that

$$\|I - H_e H\|_\infty < 1.$$

However, without *a priori* knowledge of the plant, it may not be possible to choose H_e properly. Theorem 5.1 simply says that there exist gains that yield convergence, but does not say how to pick them. It turns out that for an LTI system, operating on a finite interval, it is not really necessary to implement the scheme as described above. Instead, a much simpler approach solves our problem in exactly one trial. Because we are restricted to a finite interval, we have $y = Hu$, where H is the Toeplitz matrix given above. As previously noted, this matrix is

50

always invertible. Further, it is easy to see that its inverse is also a lower-triangular Toeplitz matrix. Specifically, let $A = H^{-1}$, so that if $y = Hu$, then $u = Ay$, or

$$u = \begin{bmatrix} a_1 & 0 & 0 & \cdots & 0 \\ a_2 & a_1 & 0 & \cdots & 0 \\ a_3 & a_2 & a_1 & \cdots & 0 \\ \vdots & \vdots & \vdots & \ddots & \vdots \\ a_N & a_{N-1} & a_{N-2} & \cdots & a_1 \end{bmatrix} y.$$

From this it is easy to develop the following recursive equation for the a_i's, given u and y:

$$a_i = \frac{1}{y(1)} (u(i) - \sum_{j=1}^{i-1} a_j y(i - j + 1)).$$

If we know these parameters, we can then use A as an optimal prefilter. Simply let $u^* = Ay_d$. Then, because $A = H^{-1}$, the output of the system is given by

$$y^* = Hu^* = HAy_d = y_d.$$

Hence, we can imagine a scenario in which we take one pass through the system, recording the input to the system and the resulting output. We then compute the parameters a_i as indicated, and form the matrix A. This is then used to compute the optimal input u^*. This procedure will be illustrated with an example following the next section.

5.3 LEARNING CONTROL WITH MULTIRATE SAMPLING

One difficulty with the one-step convergence technique is that it can produce control signals $u^*(t)$ that become unbounded with time. This will happen when the plant is non-minimum phase (or unstable, although we have assumed that the systems we are working with are either stable or have been stabilized). In this section we give an approach to learning control for non-minimum phase systems. Implementation of the method requires a multirate controller. For a summary of the necessary results from multirate sampling the reader is referred to Appendix A.

Suppose we implement a 2-delay input control scheme as described in Appendix A. This is illustrated in Figure 5.2. Figure 5.2(a) shows that we sample

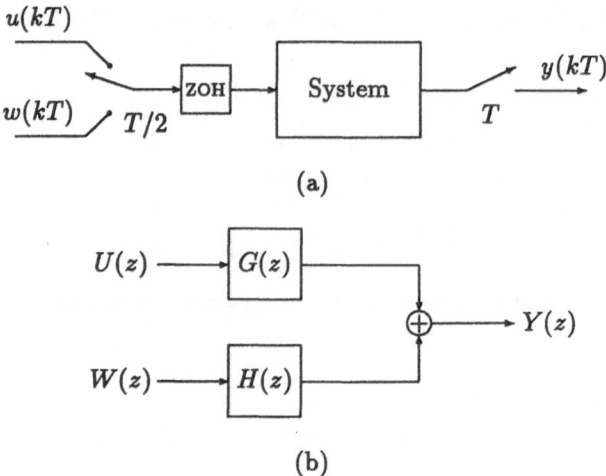

(a)

(b)

Figure 5.2: Two-delay input control: (a) sampling scheme; (b) equivalent z-transform model.

the input to the plant twice as fast as the output is sampled. In this case the resulting discrete-time system can be modelled as

$$
\begin{aligned}
x_{i+1} &= A x_i + (B_1 - B_2) u_i + B_2 w_i, \\
y_i &= C x_i.
\end{aligned}
$$

Now, define

$$
\begin{aligned}
H(z) &= C(zI - A)^{-1}(B_1 - B_2), \\
G(z) &= C(zI - A)^{-1} B_2.
\end{aligned}
$$

This gives a z-transform model of the system as shown in Figure 5.2(b). The key fact that we can use from Appendix A is that $H(z)$ and $G(z)$ have no common zeros.

To use this to our advantage, note that if we define the vectors

$$
\begin{aligned}
u &= (u(0), u(1), \cdots, u(N-1)), \\
w &= (w(0), w(1), \cdots, w(N-1)), \\
y &= (y(m), y(m+1), \cdots, y(N-1+m)),
\end{aligned}
$$

52

then we can write

$$
y = \begin{bmatrix}
h_1 & 0 & 0 & \cdots & 0 & g_1 & 0 & 0 & \cdots & 0 \\
h_2 & h_1 & 0 & \cdots & 0 & g_2 & g_1 & 0 & \cdots & 0 \\
h_3 & h_2 & h_1 & \cdots & 0 & g_3 & g_2 & g_1 & \cdots & 0 \\
\vdots & \vdots & \vdots & \ddots & \vdots & \vdots & \vdots & \vdots & \ddots & \vdots \\
h_N & h_{N-1} & h_{N-2} & \cdots & h_1 & g_N & g_{N-1} & g_{N-2} & \cdots & g_1
\end{bmatrix} \begin{pmatrix} u \\ w \end{pmatrix}
$$

$$
= P \begin{pmatrix} u \\ w \end{pmatrix}.
$$

The parameters h_i and g_i can be identified in two steps, exactly analogous to the one-step convergence case (simply let $w = 0$ to identify the h_i's and then let $u = 0$ to get the g_i's). If we then compute the Moore-Penrose pseudo-inverse [61],

$$
P^\dagger = P^T (PP^T)^{-1},
$$

the resulting optimal input (i.e., the output of the 2-delay input control based learning controller) becomes

$$
\begin{pmatrix} u^* \\ w^* \end{pmatrix} = P^\dagger y_d.
$$

This vector will be well-behaved, even when the plant is non-minimum phase and interval of interest is long. It can be shown that because the transfer functions $H(z)$ and $G(z)$ have no common zeros, the inverse of PP^T will always exist. This method is illustrated by an example in the next section.

One practical limitation of the method is an adverse effect on the intersample behavior of the output response of the system. As demonstrated in [6], this is an effect of the N-delay input multirate sampling method. This is an area for more research and, until the issues of intersample behavior are resolved, the learning control technique presented in this section may not be practically implementable.

5.4 EXAMPLES

In this section we present two examples. The first example considers the use of the one-step convergence method for a DC-motor control system. The second example shows the 2-delay input control technique applied to learning control.

5.4.1 DC-Motor

Consider a DC-motor with viscous damping as shown in Figure 5.3(a) [62]. The angular velocity ω is to be controlled with a standard unity-feedback controller, using a microprocessor. After discretization, using a sampling time of $T = .01$ seconds and a nominal choice of parameters, the plant can be modeled as

$$P(z) = \frac{1.6588z - 1.359}{z^2 - 1.15188z + .5488}.$$

To ensure zero steady-state tracking of step inputs, we choose a digital controller of the form

$$C(z) = \frac{K}{z - 1}.$$

With $K = .1$ the resulting closed-loop system is stable, with poles at $z = .7837$ and $z = .8676 \pm j.3476$. Now, suppose we desire to track a square-wave type of input on the interval $[0, 1.5]$. This would be a typical input signal if the DC-motor was used to move the rotor through a prescribed angle and then rotate it back to rest. The response of the system to a unit square-wave of duration 1.5 seconds is shown in Figure 5.3(b). Note the oscillatory nature of the response due to the complex-conjugate poles near the unit circle. A root locus plot of the system is given in Figure 5.4. This shows that not much improvement can be attained by changing the gain K. Therefore we design an optimal input as follows.

First, form the vectors $r, y \in R^{150}$ from the input/output data. This gives

$$r = (1, 1, \cdots, 1, 1, -1, -1, \cdots, -1, -1)^T,$$
$$y = (0.165, 0.447, \cdots, 0.674, 0.109, -0.572, -1.25, \cdots, -1.013, -1.009)^T.$$

Next, using the formula

$$a_i = \frac{1}{y(1)}(r(i) - \sum_{j=1}^{i-1} a_j y(i - j + 1)),$$

compute the parameters of the inverse system. This results in $A = H^{-1}$ given by

$$A = \begin{bmatrix} 6.028 & 0 & 0 & 0 & \cdots & 0 \\ -10.245 & 6.028 & 0 & 0 & \cdots & 0 \\ 5.070 & -10.245 & 6.028 & 0 & \cdots & 0 \\ .0264 & 5.070 & -10.245 & 6.028 & \cdots & 0 \\ \vdots & \vdots & \vdots & \vdots & \ddots & \vdots \\ -5.35 \times 10^{-5} & 4.16 \times 10^{-5} & -2.35 \times 10^{-5} & 2.58 \times 10^{-5} & \cdots & 6.028 \end{bmatrix}.$$

54

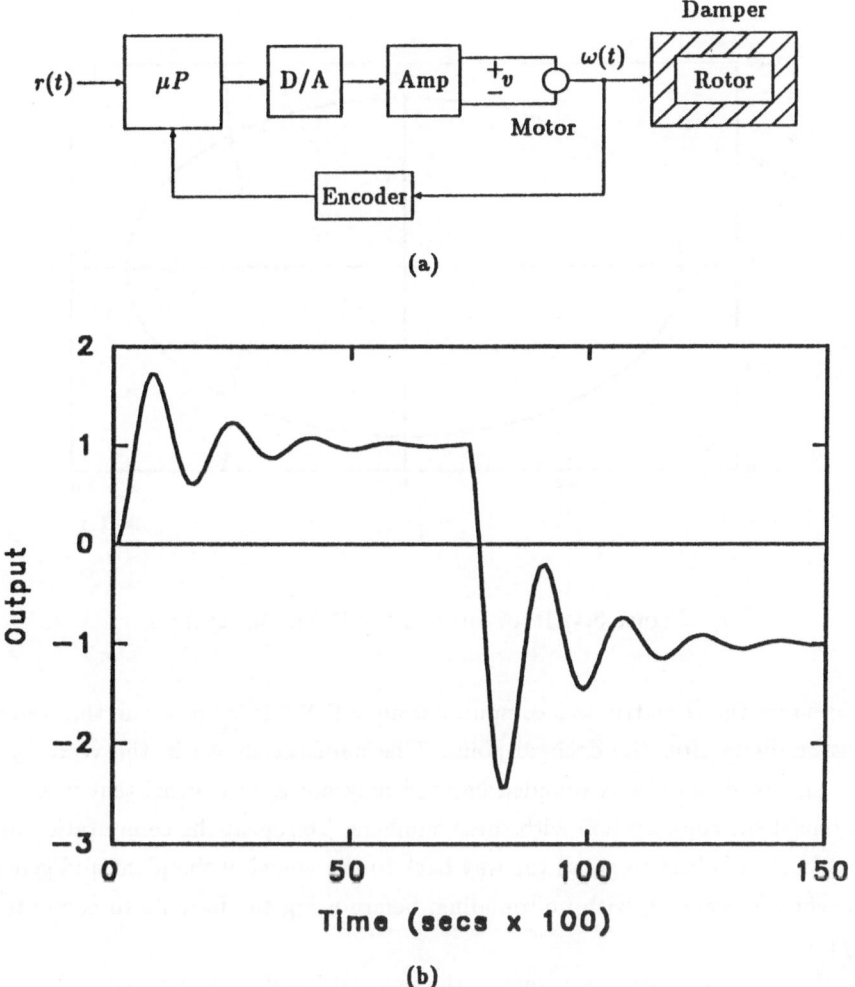

(a)

(b)

Figure 5.2: DC-motor with a viscous-inertia damper.

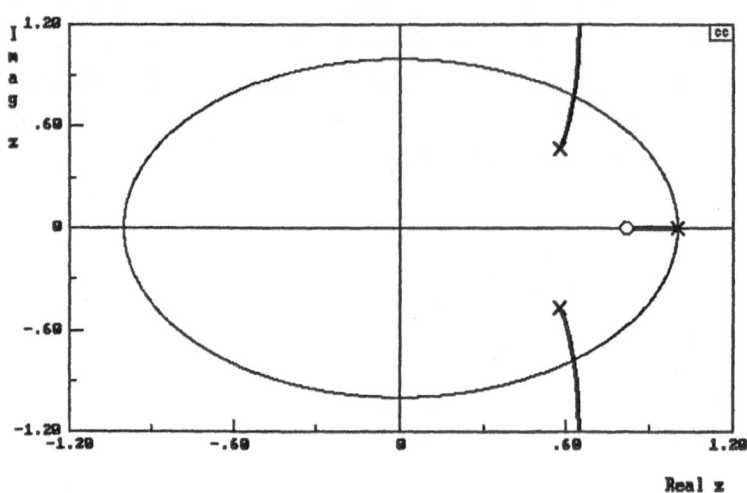

Figure 5.4: Root locus of the DC-motor system.

(Note that the A matrix was computed using a FORTRAN program that retained sixteen digits after the decimal point. The numbers shown in the vector y and the matrix A have been rounded off, and may not appear consistent if one tries to repeat the computation with these numbers. To repeat the computation of the matrix A, it is best to go all the way back to the model of the plant and generate the correct vector y, with no rounding, before using the formula to compute the a_{ij}'s.)

Next, given A, we can compute the optimal input $r^* = Ar_d$, where r_d is any desired output waveform. For the square-wave we get the optimal input sequence

$$r^* = (6.028, -4.217, .853, \cdots, 1.0, -11.056, 9.434, -.706, \cdots, -1.0, -1.0).$$

When this signal is applied to the system the resulting output is exactly the square-wave, defined on $t = .02$ to $t = 1.51$ (because of the two-step delay through the

56

closed-loop system). This is shown in Table 5.1. Also shown in Table 5.1 is the
optimal input and corresponding output when a triangular response with a unit
slope is desired. The input r^* for this case is again computed from $r^* = Ar_d$, where
r_d is the desired triangular response and A is the same matrix computed from the
first set of input/output data. As seen in Table 5.1, there is perfect tracking of
this new signal with no training of any type required. Note that, for this system,
we cannot achieve convergence with an Arimoto-style learning controller for either
of these desired responses. This is not surprising, because $|CB| \neq 0$, as required
for Arimoto et al.'s scheme. As noted in the first section of this chapter, we could
obtain convergence with a modified Arimoto-style learning control law of the form
$u_{k+1}(i) = u_k(i) + ae_k(i + 2)$ for some gain a. This would account for the two step
delay through the closed-loop system and would converge in the limit as $k \to \infty$,
if a is chosen properly. But, we comment again that such an algorithm would
not retain any knowledge of the learning. With our method, however, we are able
to generate the inverse system matrix for the finite interval with only one trial
and with no knowledge of the system dynamics. Then, after a single trial we can
generate the optimal input to the system for any desired output response, with
no further learning required.

5.4.2 Non-Minimum Phase System

To provide an example of the apparent utility of the multirate sampling technique
consider a continuous-time plant $G(s)$, with output sampling at a rate T and
input sampling at $T/2$, resulting in a basic discrete-time system (BDTS)

$$Y(z) = \frac{2(z + 2)}{z^2 - .75z + .125} U(z),$$

and a corresponding 2-delay input model

$$Y(z) = \frac{z + 2.5}{z^2 - .75z + .125} U(z) + \frac{z + 1.5}{z^2 - .75z + .125} W(z).$$

Because the BDTS is non-minimum phase, we expect the one-step convergence
method (in fact, any learning control method) to result in a control input which
gets large as the time interval of interest gets larger. For this system we carried
out the procedure of the previous example for the BDTS and also computed the
Moore-Penrose inverse associated with the 2-delay input model. The resulting
optimal input reference signal for a step response is shown for each case in Table
5.2.

57

Table 5.1: Input/Output Data: DC-Motor System

TIME	STEP		TRIANGLE	
	Optimal Input Sequence	Resulting Output Response	Optimal Input Sequence	Resulting Output Response
0.00	6.028454	0.0	0.0	0.0
0.01	-4.2173	0.0	.602	0.0
0.02	.853	1.0	.181	0.0
0.03	.879	1.0	.266	.1
0.04	.901	1.0	.354	.2
0.05	.919	1.0	.444	.3
⋮	⋮	⋮	⋮	⋮
.72	1.0	1.0	7.1	7.0
.73	1.0	1.0	7.2	7.1
.74	-11.056	1.0	7.9	7.2
.75	9.431	1.0	5.77	7.3
.76	-.706	-1.0	7.22	7.5
.77	-.759	-1.0	7.15	7.4
.78	-.803	-1.0	7.08	7.3
⋮	⋮	⋮	⋮	⋮
1.48	-1.0	-1.0	.199	.3
1.49	-1.0	-1.0	.099	.2
1.50	0.0	-1.0	.0.0	.1
1.51	0.0	-1.0	0.0	0.0

Table 5.2: Comparison of Control Signals: Non-Minimum Phase System

TIME	BDTS		2-DELAY		
	Optimal Input Sequence	Resulting Output Response	Optimal Input u_i	Optimal Input w_i	Resulting Output Response
0	0.5	0.0	—	—	—
1	1.1	1.0	−.936	1.936	0.0
2	2.4	1.0	.522	−.836	1.0
3	5.06	1.0	−.138	.461	1.0
4	10.3	1.0	.167	−.139	1.0
5	20.8	1.0	.0263	.139	1.0
6	41.8	1.0	.0916	.01	1.0
7	83.8	1.0	.0614	.0694	1.0
8	167.8	1.0	.0754	.042	1.0
9	355.8	1.0	.0689	.0547	1.0
10	671.8	1.0	.0719	.0488	1.0
11	1343.8	1.0	.0705	.0515	1.0
12	2687.8	1.0	.0711	.0503	1.0
13	5375	1.0	.0708	.0509	1.0
14	10751	1.0	.0710	.0506	1.0
15	21503	1.0	.0709	.0507	1.0
16	43007	1.0	.0709	.0507	1.0
⋮	⋮	⋮	⋮	⋮	⋮

As expected, the non-minimum phase characteristic of the basic discrete-time model results in a growing input, whereas using the 2-delay input control results in well-behaved control signals. In fact, in the 2-delay input control method the signals settle down to constant steady-state values. By changing the input twice for each output measurement, we have managed to invert the system at the sample points, without an unbounded control signal. Of course, our simulation only shows the sample point behavior. A hybrid simulation of the sampled-data system would reveal intersample oscillation that may not be acceptable in a given application. For more details about this effect the reader is again referred to [6] or [63].

5.5   COMMENTS AND EXTENSIONS

What we have done in this chapter is to present the obvious solution to the learning control problem. From chapter 3 we saw that the effect of a learning controller is to generate the output of the best approximate inverse of the plant. By discounting the delay and restricting the problem to a fixed, finite-time interval, an inverse system always exists. For the LTI case, this inverse can be computed from a single input/output measurement.

A second comment is that for the case of linear time-varying (LTV) systems we can give a development similar to that given in the first section of this chapter. For instance, for the truncated l_∞-norm we would derive a condition of the form

$$\|I - H_e H\|_\infty < 1.$$

In this case, however, the system has the form

$$H = \begin{bmatrix} h_{11} & 0 & 0 & \cdots & 0 \\ h_{21} & h_{22} & 0 & \cdots & 0 \\ h_{31} & h_{32} & h_{33} & \cdots & 0 \\ \vdots & \vdots & \vdots & \ddots & \vdots \\ h_{N1} & h_{N2} & h_{N3} & \cdots & h_{NN} \end{bmatrix},$$

and the corresponding matrix H_e would be of the form

$$H_e = \begin{bmatrix} \alpha_{11}y_d(1) & 0 & 0 & \cdots & 0 \\ \alpha_{21}y_d(2) & \alpha_{22}y_d(1) & 0 & \cdots & 0 \\ \alpha_{31}y_d(3) & \alpha_{32}y_d(2) & \alpha_{33}y_d(1) & \cdots & 0 \\ \vdots & \vdots & \vdots & \ddots & \vdots \\ \alpha_{N1}y_d(N) & \alpha_{N2}y_d(N-1) & \alpha_{N3}y_d(N-2) & \cdots & \alpha_{NN}y_d(1) \end{bmatrix}.$$

In general, for the learning controller to invert the LTV system it will be necessary to pick $N^2/2$ parameters in such a way as to ensure convergence. Note that the resulting learning controller is itself a time-varying system (this is the same implication that arises from a result on LTV systems given by Arimoto et al. [9]). The unanswered question is how these parameters are to be picked. Some type of gradient search might be employed, but as N gets large this may become impractical. Another approach is to use the fact that we are operating on a finite interval of length N. This implies that the system can be exactly identified on this interval in N trials. To see this, consider a series of N trials where the N inputs u_k are independent. We then solve the equation

$$[y_1 y_2 \cdots y_N] = H[u_1 u_2 \cdots u_N].$$

The u_i's could be chosen so that the matrix $[u_1 u_2 \cdots u_N]$ is easily inverted (make it Toeplitz, for example). Unfortunately, such an idea could only work for small N, because it may not be feasible to run a large number of training trials. Further study is needed to characterize those classes of LTV systems for which it is possible to ensure convergence with a smaller number of free parameters.

The main conclusion of this chapter is that for LTI, finite-horizon problems, there is no need to perform repeated operations to improve the tracking error of the system. This complements the conclusions of the previous two chapters. We note that similar conclusions do not apply for the LTV case or for the nonlinear case. In the next chapter we discuss the application of learning control to nonlinear plants.

CHAPTER 6

NONLINEAR LEARNING CONTROL

From the results of the past three chapters, it is clear that for a linear plant there is no real advantage to the method of iterative learning control (in the presence of noise this claim may not be as strong, however, this is still an area of research). The real usefulness of learning control is for situations in which we wish to control the performance of a nonlinear and/or time-varying system. In this chapter we consider nonlinear learning control. The first section discusses the issues involved. In the second section we give an example of a specific learning controller for a class of nonlinear systems that includes the models of typical n-jointed robotic manipulators. This learning controller has a linear time-varying structure and is illustrated with the results of a simulation experiment.

6.1 LEARNING CONTROL FOR NONLINEAR SYSTEMS

Suppose now that our system is defined by

$$y_k(t) = f_P(u_k(t), t),$$

where f_P is a nonlinear function. According to our problem statement from Chapter 2, we seek a (possibly nonlinear) system L such that the sequence

$$u_{k+1}(t) = f_L(u_k(t), y_k(t), y_d(t), t) \rightarrow u^*(t),$$

where the optimal input $u^*(t)$ minimizes the norm of the final error

$$\|y_d(t) - f_P(u^*(t), t)\|.$$

To deal with this problem, we can proceed as in Chapter 3 and use a contraction mapping requirement to develop sufficient conditions for convergence. This is the approach of almost every result available in the literature that deals with nonlinear learning control. Suppose that we have $u_k(t) \in U$, where U is an appropriately defined space. Then for the learning algorithm

$$u_{k+1} = f_L(u_k, f_P(u_k), y_d),$$

we will obtain convergence if for all $x, y \in U$ there exists a constant $0 < \rho < 1$ so that

$$\|f_L(x, f_P(x), y_d) - f_L(y, f_P(y), y_d)\| \leq \rho\|x - y\|.$$

The question that can then be posed is the following: what conditions on the plant and the learning controller will ensure that the iteration is a contraction mapping?

To establish conditions under which the learning control system is a contraction requires that we impose various levels of assumptions on either our plant and/or our learning controller. The various results for nonlinear learning control available in the literature can be distinguished in some measure by the different restrictions that are placed on the system. Once we establish general features of the system and the learning controller, we can then analyze the resulting composite system from the perspective of a contraction mapping. Three examples of the results that can be attained using this approach are given next.

In [49], Wang proposes a learning controller of the form

$$u_{k+1} = u_k + (y_d - y_k).$$

This is a simple linear learning control algorithm, with unity weighting on both the current input and the current error. If $y_k(t) = f_P(u_k)$, where $f_P(\cdot)$ is continuous on the interval of interest, then convergence is guaranteed if

$$\|I - f_P'(u)\| < 1$$

for all inputs $u \in S$, where S is convex subset of the space of continuous function and $f_P'(u)$ is the derivative of f_P with respect to its argument u. Wang's result thus defines a class of systems for which the learning control algorithm will converge. Notice that the algorithm does not need to know any information about the plant. We only require that a function of its derivative be bounded.

Wang's result is very general and is not always easily checked. It is possible to obtain less general results by restricting the plant and the learning controller to have a more specific structure. For example, Sugie and Ono provide the following result in [53]. Let the plant be described by the time-varying nonlinear system

$$
\begin{aligned}
\dot{x}_k &= a(x_k, t) + b_p(t)u_k, \\
y_k &= c(x_k, t) + d_p(t)u_k,
\end{aligned}
$$

with $a(x, t)$ and $c(x, t)$ Lipschitz in their arguments. Also, let the learning controller be given by the linear time-varying system

$$
\begin{aligned}
\dot{v}_k &= A_c(t)v_k + B_c(t)e_k, \\
u_{k+1} &= C_c(t)x_k + D_c(t)e_k + u_k.
\end{aligned}
$$

For this setup, they show convergence if

$$
\|I - d_p(t)D_c(t)\| < 1.
$$

Notice that this scheme only depends on the direct transmission terms from the plant and the learning controller. If either $d_p(t)$ or $D_c(t)$ are zero, then the condition is not satisfied. As noted in [53], this provides another explanation of the need for derivative action in Arimoto et al.'s original learning control scheme (because we can recover the effect of a transmission by differentiating the output).

Sugie and Ono's result is similar to an earlier result by Hauser [51]. For the system

$$
\begin{aligned}
\dot{x}_k &= f(x_k, t) + B(x, t)u_k, \\
y_k(t) &= g(x_k, t),
\end{aligned}
$$

with a learning algorithm given by

$$
u_{k+1} = u_k + L(y_k)(\ddot{y}_d - \ddot{y}_k),
$$

he showed convergence if $f(\cdot)$ and $B(\cdot)$ are Lipschitz and $L(\cdot)$ satisfies

$$
\|I - L(g(x, t), t)g(x, t)B(x, t)\| < 1.
$$

We can see that this convergence condition has the same form as in the other examples.

An alternative to applying contraction mapping conditions is to use Lyapunov analysis. This is typical in the analysis of learning in neural networks, which can be considered a type of learning control problem. This is also the approach taken in the convergence analysis of a novel learning control scheme proposed by Messner, Horowitz, Kao, and Boals [46]. Their technique is based upon a new method of nonlinear function identification. As in the case of contraction mapping techniques, however, the use of Lyapunov analysis techniques requires that we place varying levels of assumptions on the plant and the learning controller in order to obtain useful convergence conditions.

There is not a unifying theory of iterative learning control for nonlinear systems. Results from contraction mapping or Lyapunov techniques usually provide sufficient, but general conditions for convergence, which must be applied on a case-by-case basis. To obtain more useful results it is often necessary to restrict our attention to more specific systems. One example of this is the problem of learning control for robotic manipulators. By assuming that the nonlinear system has the standard functional form typical of a manipulator, researchers have been able to establish specific learning controllers that will converge. We illustrate this, in the next section, where we present a specific learning control scheme that can be viewed as a special case of the class of nonlinear systems studied by Hauser [51].

6.2 LEARNING CONTROLLER FOR A CLASS OF NONLINEAR SYSTEMS

Applications of robotic manipulators usually involve some type of repetitive motion. Thus, it is natural to consider iterative learning control as a way to improve the performance of a manipulator. In this section we describe a learning control scheme that can be applied to n-jointed manipulators and other systems whose models fall into a particular class of nonlinear systems. This class of equations is typical of manipulators and other motion control problems (such as the models found in aerospace systems). The technique we present can be applied to such systems whenever a given task is to be performed over and over again. We develop our method by modifying a learning control algorithm given by Bondi, Casalino, and Gambardella [7]. Their method is a model-reference adaptive control approach, based on a linear output feedback control law together with a learning controller that uses linear error feedback. For their scheme they have shown the existence of feedback gains that ensure convergence of the learning control iterations. Our method introduces an adaptive gain adjustment technique to enable the system to adaptively choose the gains that yield convergence. The effectiveness of the result is illustrated with a simulation.

6.2.1 Preliminaries

Suppose we have a nonlinear system with dynamics defined by

$$A(x_k)\ddot{x}_k + B(x_k, \dot{x}_k)\dot{x}_k + C(x_k) = u_k.$$

We suppose that u_k is an n-dimensional input vector, which drives the system at the kth trial, and $x_k \in R^n$ is the resulting output vector. For instance, if u_k is a vector of joint actuator torques and x_k is a vector of the corresponding joint angles (or positions relative to some generalized coordinate space), then this would be a typical model of an n-jointed manipulator. In the following we will place assumptions on our system that are usually made for manipulators. First, we assume that the matrix A is positive definite. Second, we suppose that the matrices A and B, and the vector C are all continuous, locally Lipschitz functions of their arguments. We also assume that A, B, and C are unknown, with the following exception. Let $x_d(t)$ be the desired output trajectory on the interval $[0, T]$. Then we additionally assume that the initial condition $C(x_d(0))$ is known.

We will define our learning controller performance in terms of the position, velocity, and acceleration of the output signal. Consequently, define the vectors

$$y_k = (x_k^T, \dot{x}_k^T, \ddot{x}_k^T)^T,$$
$$y_d = (x_d^T, \dot{x}_d^T, \ddot{x}_d^T)^T,$$

to represent the complete system output at the kth trial and the complete desired trajectory, respectively. Also, let the trajectory error at the kth trial be given by $e_k = y_k - y_d$.

Figure 6.1 shows the learning controller configuration presented by Bondi et al. As shown in Figure 6.1, the system is stabilized with a conventional linear output feedback control law of the form

$$u_k = r_k - \alpha \Gamma y_k + C(x_d(0)).$$

Here $\Gamma \in R^{n \times 3n}$ is the feedback gain matrix. Actually, we can think of this as either state feedback or output feedback. In point of fact, the vector $y_k(t)$ will

67

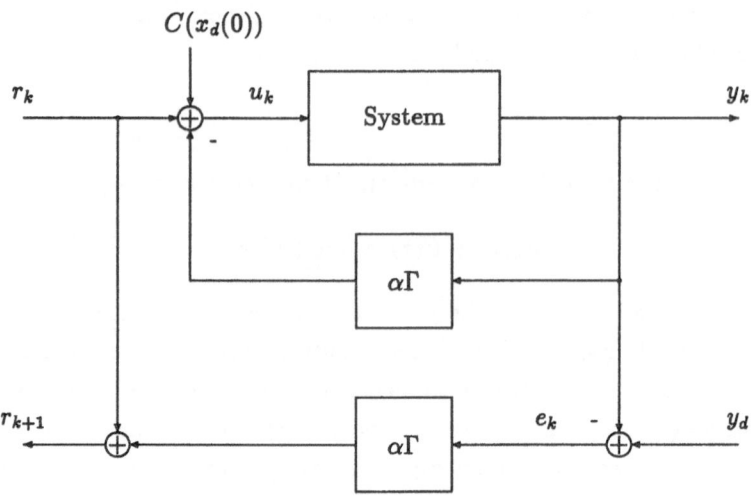

$C(x_d(0))$

r_k u_k System y_k

$\alpha\Gamma$

r_{k+1} $\alpha\Gamma$ e_k y_d

Figure 6.1: High-gain feedback learning control scheme.

typically contain all the states of the system. However, if we view y_k as the output, then the control law can be thought of as implementing output feedback.

The other feature of Figure 6.1 is the reference input r_k. This signal can be described as a time-varying reference input. It is updated by the learning control law

$$r_{k+1} = r_k + \alpha\Gamma e_k.$$

This algorithm uses the same feedback gain that is used to stabilize the system.

The operation of the system is the same as for any learning control system. A reference input r_0 is chosen to drive the manipulator during the first trial. At the end of each trial, the current reference input r_k and the resulting output error e_k are used to compute the reference r_{k+1} for the next trial. As in previous schemes, the initial conditions are reset at the beginning of each trial. Bondi et al. were able to prove the following convergence result for this learning control scheme.

Theorem 6.1 *([7] Theorem 2) Let the feedback matrix Γ be partitioned as $\Gamma = [P, L, K]$, with P, L, and K nonsingular and $\det(Ks^2 + Ls + P)$ Hurwitz. Then* *there exists a lower bound $\bar{\alpha} > 0$ so that if $\alpha \geq \bar{\alpha}$, then the feedback gain $\alpha\Gamma$ ensures that for all $t \in [0, T]$ we have*

$$\lim_{k \to \infty} \|e_k\| = 0.$$

This theorem says that there exists a high-gain learning controller that will cause the error to converge. However, in order to use the results of this theorem we must know $\bar{\alpha}$ *a priori*. To avoid this, we propose an adaptive method for obtaining gains large enough to guarantee convergence. Our method will modify the configuration of Figure 6.1 by introducing time-varying gains α_k that are updated at each trial. Before we can proceed, however, we need the following corollary to Theorem 6.1.

Corollary 6.1 *Let the gain matrix $\alpha\Gamma$ in Figure 6.1 be replaced by the time-varying gain matrix $\alpha_k\Gamma$, where Γ is as in Theorem 6.1 Also, let $\bar{\alpha}$ be the minimum gain required for convergence from Theorem 6.1. If the sequence $\{\alpha_k\}$ is monotone increasing with $\alpha_0 \geq \bar{\alpha}$, then for all $t \in [0, T]$ we have*

$$\lim_{k \to \infty} \|e_k\| = 0.$$

The proof of this Corollary follows from an examination of the proof of Theorem 6.1. Rather than reproduce the proof we describe the essential features and refer the interested reader to [7] for more details. There are two parts to the proof. First, it is demonstrated that the uniform boundedness of the trajectory error at each trial can be guaranteed, given a gain α greater than $\bar{\alpha}$. The argument of this part of the proof is easily extended to the case of an increasing sequence of α's as described in the Corollary 6.1. Second, using the uniform boundedness property and the other assumptions on the system, it is shown that there exists a Lipschitz constant $\rho(\alpha) < 1$ that controls the convergence of the position error. This fact leads to Theorem 6.1. Now, this Lipschitz constant is monotonically decreasing toward zero for increasing α, which implies that for a monotonically increasing sequence $\{\alpha_k\}$ we have a monotonically decreasing sequence of Lipschitz constants satisfying $\rho(\alpha_{k+1}) \leq \rho(\alpha_k)$. From this the convergence result of the corollary follows.

6.2.2 Adaptive Gain Adjustment

In Theorem 6.1 we could ensure convergence by using a gain $\alpha \geq \bar{\alpha}$ in the learning control configuration. However, this convergence result does not tell us how to obtain the correct value of $\bar{\alpha}$. One solution would be to simply try larger and larger values of α until convergence is obtained. Alternately we can develop an adaptive scheme that automatically increases α until the convergence property is attained. In this subsection we will describe such an adaptive technique. The procedure is shown in Figure 6.2. In this figure the feedback gain matrix is now

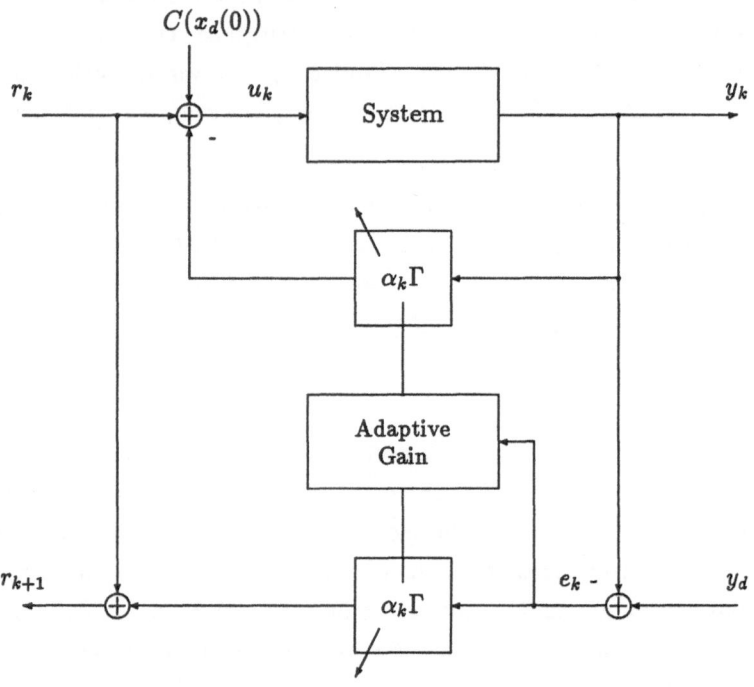

Figure 6.2: Adaptive gain adjustment configuration.

a time-varying feedback gain $\alpha_k\Gamma$. Thus we have put a time-varying controller on the system, instead of the constant gain controller used in the previous subsection. Also, the learning control law is now given by

$$r_{k+1} = r_k + \alpha_k\Gamma e_k.$$

The gain α_k is updated at each trial according to

$$\alpha_{k+1} = \alpha_k + \|e_k\|^m,$$

for $m > 0$. This gives us a time-varying learning controller. The basic idea is to make α_k larger at each trial by adding some positive number. In this way it will eventually become bigger than $\bar{\alpha}$. When this happens the results of Corollary 6.1 can be applied to the successive iterations and we will have a convergent learning control scheme. Notice that at each trial we add the norm of the error. This way, when the algorithm begins to converge, the gain α_k will eventually stop growing, because the norm of the error will go to zero. The following theorem formalizes the result.

Theorem 6.2 *Let the learning control configuration shown in Figure 6.2 satisfy the previous assumptions on the system and let*

(a) $r_{k+1} = r_k + \alpha_k \Gamma e_k$.

(b) $u_k = r_k + \alpha_k \Gamma y_k$, *where Γ is as in Theorem 6.1.*

(c) $\alpha_{k+1} = \alpha_k + \|e_k\|^m$, *for $m > 0$.*

Then for all $t \in [0, T]$,

$$\lim_{k \to \infty} \|e_k\| = 0.$$

Proof: For simplicity let $m = 1$. This causes no loss of generality. We will argue by contradiction. Hence, suppose that the error does not converge. That is, let $\lim_{k \to \infty} \|e_k\| \neq 0$. Then we have two cases to consider.

(i) $\limsup_{k \to \infty} \|e_k\| = \infty$. In this case there exists a sequence $\{k_i\}$ such that

$$\|e_{k_{i+1}}\| > \|e_{k_i}\| > 1,$$

for $i = 1, 2, \ldots$. Then

$$\alpha_{k_i+1} = \alpha_{k_i} + \|e_{k_i}\| > \alpha_{k_i} + 1,$$

so the subsequence $\{\alpha_{k_i}\}$ diverges as $i \to \infty$ and hence, $\lim_{k \to \infty} \alpha_k = \infty$.

(ii) $\limsup_{k \to \infty} \|e_k\| = C > 0$, where C is a constant. In this case there exists a sequence $\{k_i\}$ such that $\lim_{i \to \infty} \|e_{k_i}\| = C$. Thus there exists an integer N such that $\|e_{k_i}\| > C/2$ whenever $i > N$. But, when $i > N$,

$$\alpha_{k_i+1} = \alpha_{k_i} + \|e_{k_i}\| > \alpha_{k_i} + C/2,$$

so the subsequence $\{\alpha_{k_i}\}$ diverges as $i \to \infty$ and hence, $\lim_{k \to \infty} \alpha_k = \infty$.

Thus, if the error does not converge, then we can conclude that the sequence of gains α_k is increasing without bound. This means that there must exist an integer $\bar{k}$ such that $\alpha_{\bar{k}} \geq \bar{\alpha}$. But, our update equation ensures that $\alpha_{k+1} \geq \alpha_k$ for all k, because $\|e_k\| \geq 0$. So we have $\alpha_{k+1} \geq \alpha_k \geq \bar{\alpha}$ for all $k > \bar{k}$ and by Corollary 6.1 this implies that $\lim_{k \to \infty} \|e_k\| = 0$. However, this contradicts our assumption. **QED**

There are a number of issues to consider when applying Theorem 6.2. One question is which norm to use. By using different norms of the error, one could develop any number of different learning control schemes. For instance, we could design a completely adaptive law that changes the gain by using a standard Euclidean norm defined at each instance of time in the interval $[0, T]$. That is, let

$$\alpha_{k+1}(t) = \alpha_k(t) + \|e_k(t)\|^m.$$

In this case we have blurred the difference between a learning controller and a conventional adaptive controller. A second approach would be to keep the gain constant during each trial, updating it only at the end of the trial, with a finite record of the error. An example of this would be to let

$$\alpha_{k+1} = \alpha_k + \|e_k(T)\|^m$$

(again using the Euclidean norm). This is the type of norm used in the example below. Another approach would be to use some type of cumulative measure of the error over the entire trial. For instance, we could use

$$\alpha_{k+1} = \alpha_k + \|e_k\|_{L_p[0,T]}^m,$$

where we use a standard L_p-norm.

A second issue is the rate of convergence. Notice that the parameter m in our gain adjustment is not really needed. As noted in the proof of Theorem 6.2, we can always choose $m = 1$. However, we can influence the convergence rate through choice of this parameter in the gain adjustment law. We would like α_k to increase rapidly at first, but after $\alpha_k > \bar{\alpha}$ we would want α_k to stop increasing (so we will not drive the system or its actuators to saturation). By making m large at first, but then reducing it we can shape the convergence properties of the algorithm. This is an area for further study.

On a final note, we point out a learning control method reported by Kawamura, Miyazaki, and Arimoto [21] that is related to Bondi et al.'s scheme. Kawamura et al. consider a learning control law of the form

$$r_{k+1} = r_k + \bar{\alpha}\Lambda e_k.$$

Similar to Theorem 6.1, they show that convergence can be guaranteed if $\bar{\alpha}$ is large enough. For this learning control scheme it is possible to give a result analogous to our Theorem 6.2 by utilizing an adaptive gain adjustment algorithm.

6.2.3 Simulation Experiment

We have applied the learning control algorithm described in the previous subsection to a two-joint manipulator. The system is shown in Figure 6.3. To model the manipulator, let $x(t) = (\theta_1(t), \theta_2(t))^T$ [40]. Then, for the masses and lengths shown in Figure 6.3, we can describe the system dynamics as

$$A(x_k)\ddot{x}_k + B(x_k, \dot{x}_k)\dot{x}_k + C(x_k) = u_k,$$

with

$$
\begin{aligned}
A(x) &= \begin{pmatrix} .54 + .27\cos\theta_2 & .135 + .135\cos\theta_2 \\ .135 + .135\cos\theta_2 & .135 \end{pmatrix}, \\
B(x, \dot{x}) &= \begin{pmatrix} .135\sin\theta_2 & 0 \\ -.27\sin\theta_2 & -.135(\sin\theta_2)\dot{\theta}_2 \end{pmatrix}, \\
C(x) &= \begin{pmatrix} 13.1625\sin\theta_1 + 4.3875\sin(\theta_1 + \theta_2) \\ 4.3875\sin(\theta_1 + \theta_2) \end{pmatrix}.
\end{aligned}
$$

Our simulation of the learning control scheme for this system had the following characteristics:

(a) A fourth-order Runga-Kutta subroutine was used to simulate the system dynamics.

(b) Following [7], we used first-order highpass filters of the form

$$H(s) = \frac{10s}{s + 10}$$

to estimate the joint accelerations $\ddot{\theta}_1$ and $\ddot{\theta}_2$ from the joint velocities $\dot{\theta}_1$ and $\dot{\theta}_2$, respectively.

73

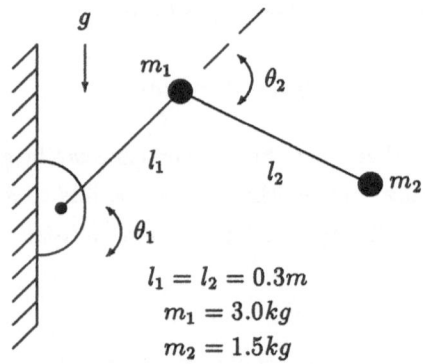

Figure 6.3: Two-joint manipulator.

(c) The gain matrices P, L, and K used in the feedback and learning control law were chosen to be

$$P = \begin{pmatrix} 50.0 & 0 \\ 0 & 50.0 \end{pmatrix}, L = \begin{pmatrix} 65.0 & 0 \\ 0 & 65.0 \end{pmatrix}, K = \begin{pmatrix} 2.5 & 0 \\ 0 & 2.5 \end{pmatrix}.$$

These matrices satisfy the Hurwitz condition specified in Theorem 6.1.

(d) To implement the learning controller with adaptive gain adjustment we updated the gain by the rule

$$\alpha_{k+1} = \alpha_k + .1\|e_k(T)\|$$

where the standard Euclidean norm is evaluated at the end point of the trial.

(e) We initialized α_0 to 0.01.

(f) The desired trajectory that we wanted to track was a triangular signal applied to each joint. Such a signal would correspond to commanding the arm to move from a starting point out to a point in the distance and then back again.

Figure 6.4 shows the resulting trajectory of θ_1 for several trials. It also shows the desired trajectory. Note how large the observed error is on the first trial. This is to be expected, because we have no knowledge of the system dynamics, other

74

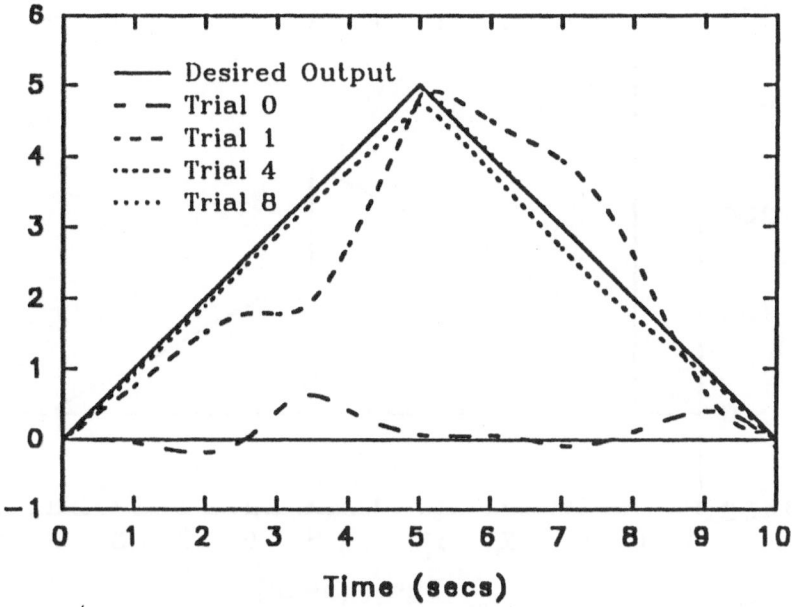

Figure 6.4: Output response of the position θ_1.

than the initial condition. However, after only one iteration the improvement is quite impressive, with the output looking distinctly triangular. As learning continues, the actual output moves closer and closer to the desired output. By the eighth trial the system is tracking the desired response with essentially no error. In Figure 6.4 it is not even possible to distinguish between the desired output and the actual output on the eighth trial. This is also true of the other signals in the vector y. It is interesting to note that without the adaptive gain adjustment, the learning scheme does not converge for the initial value $\alpha_0 = 0.01$. That is, if we run the original scheme of Bondi et al., the system will not converge with $\alpha = .01$. Thus the adaptive gain adjustment solves the problem of finding a suitable value of α *a priori*.

An important point regarding the learning control algorithm is made in Figure 6.5, which shows the reference input r_{θ_1} at the eighth trial. This is the torque

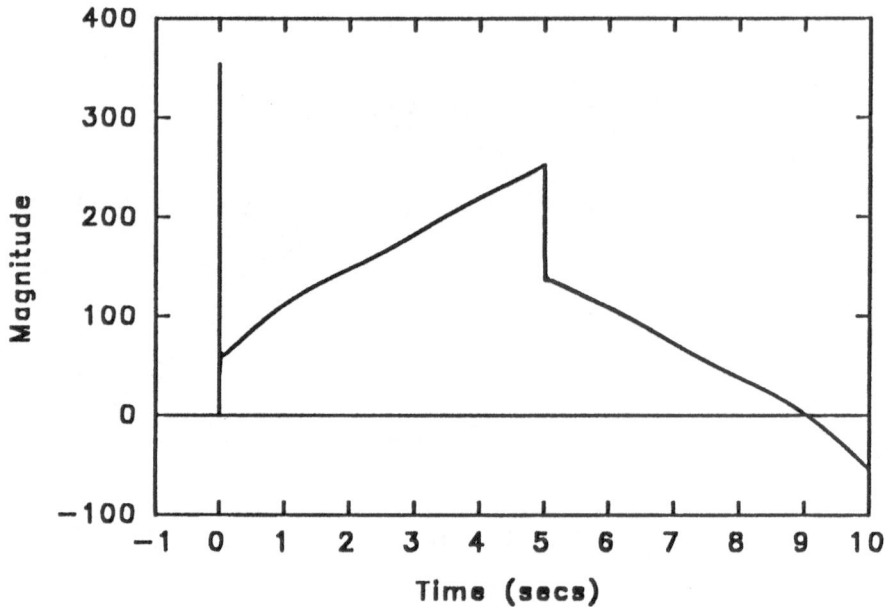

Figure 6.5: Reference input at the eighth trial.

signal that drives the system at the eighth trial, resulting in the tracking property
demonstrated in Figure 6.4. It can be seen that this signal has a very large
magnitude, both at the initial time and at the time when the trajectory turns
around. Intuitively it is clear that such large magnitude inputs are necessary
to overcome the system dynamics and obtain perfect tracking. Unfortunately, a
reference input such as this will likely cause the system actuators to saturate. If
this happens, then the system may not converge as expected. To deal with this,
it is necessary to repeat our analysis in the presence of limiting constraints on the
system inputs. More work must be done to see how such constraints will affect
the convergence properties of the learning controller. It turns out that for linear
time-invariant systems we can still obtain convergence when we bound the control
signal. However, it is not clear what the effect will be when dealing with nonlinear
systems. Also, even for the linear case, although we obtain convergence when we

place limits on our control signal, it is not clear how large this will make the norm of the converged error signal. The design question, which is so far unanswered, is how to design the learning controller so that we have both convergence and a small error norm when we have control signal constraints.

CHAPTER 7

ARTIFICIAL NEURAL NETWORKS FOR ITERATIVE LEARNING CONTROL

As we have noted, the real usefulness of iterative learning control is for problems in which we wish to control the transient response behavior of nonlinear or time-varying systems. In this case it makes sense to consider learning controllers that also have a nonlinear or time-varying structure, such as the learning control scheme demonstrated in the previous chapter. A non-trivial question, however, is what type of nonlinear system should be considered. The class of all nonlinear systems is very large and it is not clear what structure would work best for learning control applications. One answer to this question is to consider the class of nonlinear systems called artificial neural networks. Artificial neural networks, with their nonlinear structure and their ability to learn internal representations and recall associations, are good candidates for nonlinear learning control.

In this chapter we will consider some ways to use artificial neural networks (ANNs for short) in iterative learning control schemes. We begin the chapter with a discussion about the use of ANNs in control systems. We then present three different techniques for developing ANN-based learning controllers. Each method exploits the ability of a neural network to implement a nonlinear mapping. The first two approaches use feedforward nets with backpropagation learning. The first method can be thought of as a static learning control system, which achieves the transient response control by performing a pattern recognition task. The second method uses a neural net as a dynamical system. The third technique uses a different type of learning paradigm for training the ANN. The method is based on reinforcement learning and implements a stochastic automata through its control strategy.

We assume throughout that the reader is familiar with the field of ANNs. If not, Appendix B has been written as a self-contained tutorial that should provide sufficient background to understand the material in this chapter.

7.1 NEURAL NETWORK CONTROLLERS

In this section we give a brief overview of how neural nets can be used in control systems. For more details we refer the reader to a comprehensive paper by Narendra and Parthasarathy [64] and to a recent book titled *Neural Networks for Control* [65]. See also a technical report by the author on this subject [66].

Most of the applications of artificial neural networks can be grouped into four categories: (1) pattern classification and associative memory; (2) self-organizing systems; (3) solution of optimization problems; and (4) implementation of nonlinear mappings. It is important to understand how each of these ANN capabilities might be used in a control system.

Applications of the results involving pattern recognition and associative memory to control problems are difficult, because they deal with static patterns, while most control problems consider the temporal characteristics of dynamic systems. One way we might use an associative memory for control is to store control laws that change at different operating points. Suppose we have a nonlinear system that must operate at two different set points. Typically such systems would be linearized around each set point and a different control law derived for each case. Then some type of logic or rule is used to switch from one control law to another, depending on the system's operating region. A neural net can be used to implement this kind of switching rule by storing the correct control law in an associative memory (for example, see [67] and [68]). This use of an associative memory has also been proposed for storing the bang-bang control laws that often arise in optimal control theory. It is interesting to note that the same equations that arise from a perceptron-based associative memory were also given in the early 1970's by Tsypkin [22,23]. However, Tsypkin derived his equations independent of any notion of neural nets. Another approach to using associative memories for control has been given in [69]. Here the idea is to store in a neural net the correct control laws corresponding to a large number of different second-order plants. The net is used in conjunction with a feature extractor, which identifies the parameters of a second-order model of the system. This information is then passed to the neural net, which produces the corresponding controller parameters and passes them to the controller. This is essentially a gain scheduling approach, which uses a neural net to store the gains.

Two other categories of neural network properties are self-organization [70, 71] and optimization [72,73]. Self-organizing neural nets, which learn without a teacher, have been shown to be able to extract features from input data, but the potential for applying these types of algorithms to control problems is not clear, especially when dealing with deterministic systems (however, the idea of self-organizing control for stochastic systems has been developed [27]). To apply the results in optimization it is usually necessary to formulate your problem in special ways that vary from problem to problem. These results are perhaps best used in a design phase, rather than for real-time control, and the weights must be set *a priori*. More research is necessary to incorporate these ideas into useful tools for control system development, although some initial results have been developed by the author and his colleagues in the area of optimal control theory [74].

It is possible to identify a fourth category of neural net applications, based on the ability of an ANN to generate a nonlinear mapping from input space to output space (or, in the case of a recursive net, a trajectory from input to output space). This ability can be interpreted as a generalized form of an associative memory or pattern classification property. The use of an ANN as a nonlinear mapping is motivated by a result of Kolmogorov [75,76], which tells us that it is possible (with certain restrictions, of course) to implement any nonlinear mapping from R^n to R^m using a neural network. Unfortunately, this is an existence result and it does not tell us if, or when, it is possible to learn a given nonlinear mapping. However, there has been a good deal of success at actually getting neural nets to identify nonlinear systems and it is this use of ANNs that seems to have the most potential for control applications.

The ability to generate a static nonlinear mapping is not in itself a useful property for control, because most control systems operate dynamically. Hence it is necessary to generate a dynamical system with the net in order to have any real value for control applications. Recently, Narendra and Parthasarathy have suggested a way to use feedforward ANNs to emulate nonlinear dynamical systems [64]. To illustrate their idea, consider a nonlinear discrete-time dynamical system described by the equation

$$y_{k+1} = f(y_k, y_{k-1}, y_{k-2}, u_k, u_{k-1}).$$

In this system we see that the next output is a function of the current input, the current output, the most recent past input, and the two most recent past outputs. Following the approach of [64], we could imagine the configuration shown in Figure 7.1, where the operator z^{-1} denotes a unit time delay and the neural net implements the nonlinear mapping f. This configuration allows us to account for

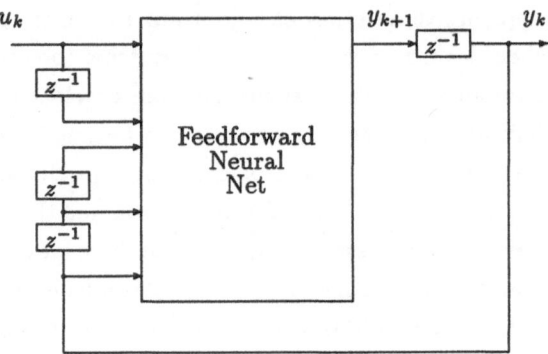

Figure 7.1: Neural net implementation of a discrete-time system.

the temporal or dynamic characteristics of the system, using a feedforward neural network to implement the nonlinearity. We will use this concept below in one of our applications of ANNs to the learning control problem.

Using an ANN in the configuration shown in Figure 7.1 suggests several approaches to control. One approach is to simply use a neural net implementation of a dynamical system as the control block in a conventional feedback control system. In the sense of a conventional control scheme, one can immediately think of three ways to exploit the ability of a neural net to implement a nonlinear mapping. Figure 7.2 shows a standard unity feedback configuration. In this configuration the

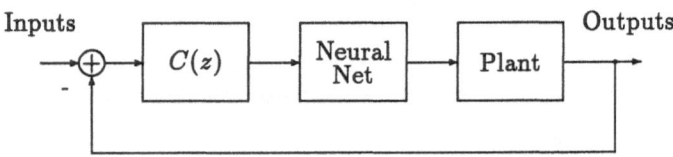

Figure 7.2: Closed-loop control.

neural net is to implement the inverse mapping of the system. Then the closed-loop dynamics would be described by the linear transfer matrix $C(z)(I + C(z))^{-1}$ Notice that in Figure 7.2 the neural net combined with the linear controller $C(z)$

acts like a nonlinear controller, with closed-loop feedback an essential element of the design. Other ideas include using the nonlinear mapping of the neural net to implement a cascaded open-loop controller (Figure 7.3) or a feedforward controller (Figure 7.4). Discussion of these themes can be found in [77]. The main idea is

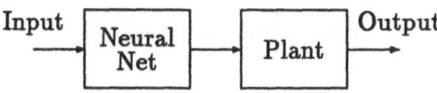

Figure 7.3: Open-loop control.

that, if we want to control a nonlinear system, then it makes sense to consider a controller that has a nonlinear structure. A feedforward neural network gives us one particular framework for such a nonlinear controller structure.

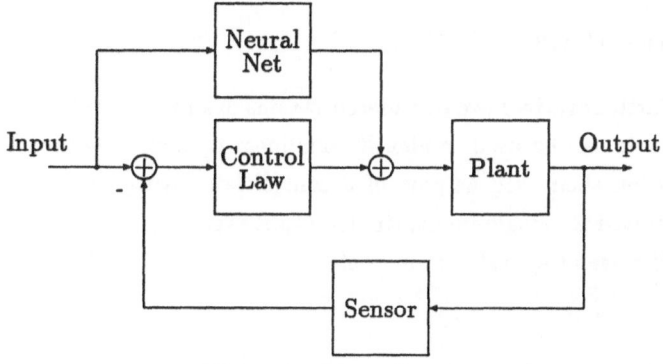

Figure 7.4: Feedforward control.

A second approach to using ANNs for control applications is to use them as components in an adaptive control scheme. Narendra and Parthasarathy have presented a preliminary analysis of the use of neural networks within the framework of conventional model reference adaptive control (MRAC) [64]. Consider, for instance, a nonlinear system (example taken from Narendra and Parthasarathy's paper) of the form

$$y_{k+1} = f(y_k, y_{k-1}) + u_k,$$

where

$$f(x, y) = \frac{xy(x + 2.5)}{1 + x^2 + y^2}.$$

The goal of the control is to match the dynamics of a reference model given by

$$y_{k+1} = .6y_k + .2y_{k-1} + r_k,$$

where r_k is the reference signal. Let the control law be given by

$$u_k = -N(y_k, y_{k-1}) + .6y_k + .2y_{k-1} + r_k.$$

Here $N(x, y)$ is a feedforward neural network that is identified using backprop-agation training. It can be seen that if the net successfully learns the system dynamics, then the output behavior will approach the desired behavior, because the closed-loop system is now given by

$$y_{k+1} = f(y_k, y_{k-1}) - N(y_k, y_{k-1}) + .6y_k + .2y_{k-1} + r(k).$$

Narendra and Parthasarathy have considered various examples of this type, show-ing how neural nets can be used to identify nonlinear dynamics, with the results of the identification then used as part of a control law. Although most of this work is demonstrated for single-input, single-output systems, it gives a very good foundation for the use of neural nets in control.

7.2 STATIC LEARNING CONTROLLER USING AN ANN

A central problem in applying neural networks to control is that most neural net-works accept spatial inputs. However, in control we are usually interested in a sequence of spatial inputs (i.e., a temporal signal). The approach of Narendra and Parthasarathy [64] gives us one way to deal with identifying the dynamic behavior of input/output data. In this section we consider another way to handle tempo-rally related input/output pairs. Our method takes advantage of the fact that any practical implementation of an iterative learning control scheme for repetitive systems will operate on a finite interval. Thus we can form a single "supervector" from the data collected on a given trial. We then input this "supervector" to our net. Thus we have effectively turned the temporal signals into static or spatial signals.

84

Consider the finite-horizon learning control problem, where we wish to shape the transient response on an interval of length N. For this problem we propose the learning controller shown in Figure 7.5. The configuration is the same as the learning control schemes we have described earlier, but now the learning control algorithm is a nonlinear function, implemented by a feedforward neural network. The neural network has N input nodes and N output nodes. The input to the neural network is the vector y_d, which we define to be

$$y_d = (y_d(1), y_d(1), \cdots, y_d(N))^T.$$

The output of the neural network is the vector

$$u_{k+1} = (u_{k+1}(0), u_{k+1}(1), \cdots, u_{k+1}(N-1))^T.$$

(We assume a one-step time delay through the system, with no loss of generality.) This signal is stored in memory until the next trial, when it is applied to the system, one step at a time. Learning is accomplished by sampling the output to form the vector

$$y_k = (y_k(1), y_k(2), \cdots, y_k(N))^T,$$

and then updating the weights based on the error vector $e_k = y_d - y_k$. We use backpropagation to update the weights (see Appendix B for the exact equations of the net and the learning algorithm). Notice that we present the neural network with a single input vector per trial, even though the trial actually produced N samples. That is, we have taken a set of N one-dimensional time-varying input/output pairs and turned them into a single N-dimensional input/output pair. Thus we call this a static learning controller, because the net tries to learn the inverse system mapping by training with static input/output patterns that were formed from the original temporal data signals.

For this scheme various examples were tested. In our examples we used computer-simulated feedforward nets with two hidden layers (as described in Appendix B). The number of neurons in each hidden layer was picked to be $M = 3N$. The learning gain η and the sigmoid constant a were usually chosen in an ad hoc way for each example (in our examples the same learning gain was used for each layer of weights). To illustrate how the technique worked, consider the non-minimum phase plant

$$P(z) = \frac{z + 1.2}{z^2 + 1.5z + .375}.$$

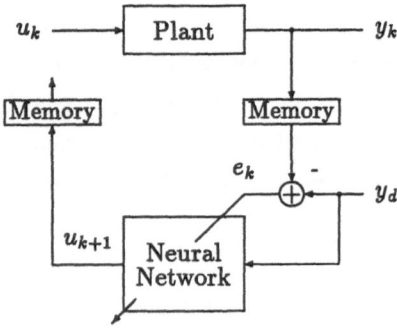

Figure 7.5: Static learning controller.

We wish to derive the input necessary to produce a unit step as the system output on the interval $[0, 20]$. Using $\eta = .1$ and $a = 1$, the scheme produced an output error whose Euclidean norm was less than .0001 after 24 iterations. Table 7.1 shows the resulting control input u^*. Also shown in Table 7.1 is the resulting optimal input u^* when the learning control scheme is applied to a nonlinear system defined by

$$y(k) = \sqrt{y(k-2)} + u^2(k-1).$$

For this example the neural network learning controller was able to produce an output error with Euclidean norm less than .005 after 54 iterations, assuming no knowledge of the system. We used $\eta = .025$ and $a = .1$ for this example. We comment that for the first example the same optimal input sequence u^* could have been found in one step, using the techniques given in the chapter on finite-horizon learning control. However, no such technique is known for the nonlinear example. This is where the real potential of learning control and, in particular, neural network-based learning control will be found.

There are a number of open issues regarding this method of using ANNs for learning control. Of particular importance is the issue of learning and convergence. For the backpropagation learning method that we used, there are no solid theoretical results on the choice of learning gains to guarantee convergence. In the examples above, a bad choice of gains would result in divergence. We also found that the learning algorithm could exhibit saddle-point behavior. For instance, the

Table 7.1: Input/Output Data: Static Learning Controller

TIME	LINEAR		NONLINEAR	
	Optimal Input Sequence	Resulting Output Response	Optimal Input Sequence	Resulting Output Response
0	1.0000	0.0000	1.0000	0.0000
1	1.3000	1.0000	0.9999	1.0000
2	1.3150	1.0000	0.0451	0.9999
3	1.2970	1.0000	0.0451	1.0020
4	1.3186	1.0000	−0.0287	1.0020
5	1.2927	1.0000	−0.0222	1.0018
6	1.3238	1.0000	−0.0048	1.0015
$\vdots$	$\vdots$	$\vdots$	$\vdots$	$\vdots$
15	1.2193	1.0000	0.0177	1.0009
16	1.4119	0.9999	0.0199	1.0008
17	1.1808	1.0000	0.0202	1.0008
18	1.4582	0.9999	0.0175	1.0008
19	1.1253	0.9999	0.0222	1.0007
20	0.0000	1.0000	0.0000	1.0009

nonlinear example we showed above would converge to a minimum error norm of about .004. However, repeated iterations after this caused the norm to begin increasing and eventually diverge. Because the nonlinearity inherent in the neural network model makes convergence analysis difficult, it is not clear how such behavior can be avoided. Another practical concern is that if N, the number of samples, becomes too large, then it becomes harder to get the learning to converge. To overcome these problems we must exploit the dynamical relationships that exist between different samples of the input and output signals. This is what is done in the next section.

7.3 DYNAMICAL LEARNING CONTROLLER USING AN ANN

In this section we use the approach of Narendra and Parthasarathy [64] to develop a nonlinear learning controller that takes into account the dynamic relationship between the input/output data. The advantage of this method over that of the previous section is that the number of inputs to the neural net of the learning controller need be no greater than twice the order of the plant. This is true no matter how large we make N, the length of the learning control trial. In contrast, with the previous method our neural net must have N inputs and N outputs. This will cause problems with convergence as N increases.

Our proposed dynamical learning control scheme is shown in Figure 7.6. We call this a dynamical learning control scheme because the final learning controller actually acts like a dynamical system, using past outputs as inputs, with the neural network used in a configuration as shown above in Figure 7.1.

Notice that in this system there are two feedforward neural networks. One, NN_2, is used to identify the forward dynamics of the plant. The other, NN_1, is used to identify the inverse dynamics of the plant. It is assumed that the order of the plant is known and that each neural net block shown in the figure is configured as in Figure 7.1 The reason for using two nets rather than just one net is to avoid the problem of projecting through a gradient of the plant to get the correct error for learning the inverse dynamics (again, see [77] for a discussion of this problem).

The idea behind this technique can be described as follows. At each trial we generate an output trajectory in response to an input trajectory. These define a set of input/output pairs that can be used to train a neural network in either the forward or reverse direction. Now, assume that two nets, NN_2 and NN_1, have been trained to produce the forward mapping and the reverse mapping, respectively,

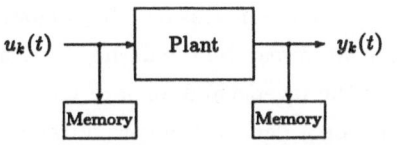

Collect I/O data
during kth trial.

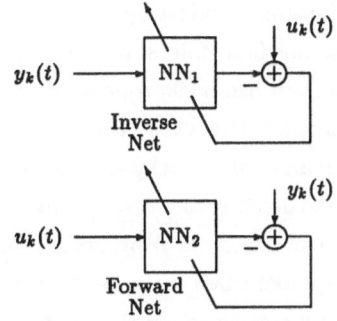

Stage 2

Train inverse and
forward nets, using
I/O data from
the kth trial.

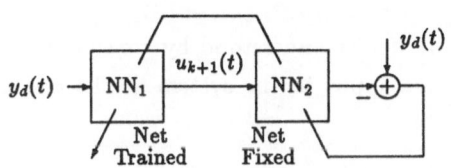

Stage 3

Compute $u_{k+1}(t)$ by
cascading NN_1 and NN_2
and adjusting NN_1 with
NN_2 held fixed.

Figure 7.6: Dynamical learning controller.

for a given set of input/output pairs. Suppose we then cascade these nets by feeding the output of the forward map NN_2 into the inverse map NN_1. Then we should be able to produce a unity map from the input of the forward map to the output of the inverse map. In the same way if we cascade the two nets in the opposite direction, by feeding the output of the inverse map NN_1 into the forward map NN_2, then we should be also able to obtain a unity mapping.

Now suppose we cascade the two nets with the inverse net NN_1 followed by the forward net NN_1. Let the input to the inverse map be the desired system output y_d. If the two nets had perfectly learned the inverse and forward mappings, then the output of the forward net NN_2 would be y_d, and the output of the inverse net NN_1 would be the optimal input required to produce y_d. Unfortunately, if the training has used only a single trajectory pair, the nets will not have implemented the correct functional mapping. Rather, they will only have learned the forward and inverse mappings in the direction of the single trajectory. Our idea to correct this is to cascade the nets (with the inverse net first, followed by the forward net) and apply the desired system output. We then use backpropagation to train the inverse net, while holding the forward net fixed. Thus we are using the forward net as a reliable model of the system dynamics and using backpropagation to project the errors from the output of the forward net back to the inverse net. This allows us to avoid having to estimate the gradient of the plant from actual data.

There is one other requirement that we make when training the cascaded nets with the forward net fixed. We also force the nets to retain the ability to map the original trajectory in both the forward direction and the inverse direction. That is, we increase the size of the training set to include the original data as well as the desired trajectory.

The ideas described above can be better understood by considering the following procedure for computing a new input to the system at the end of each trial.

Algorithm 7.1

1. *Set the initial input $u_0(t) = y_d(t)$ and set $k = 0$.*

2. *Stage 1: Run the plant with u_k. Collect and store in memory $u_k(t)$ and $y_k(t)$.*

3. *Stage 2: Parallel learning. Train the inverse and forward nets using the*

input/output data from the kth trial. Training is on a point-by-point basis, using backpropagation.

4. *Stage 3: Series learning. Cascade the inverse and forward nets. Apply the desired output $y_d(t)$ as the input to the inverse net. Train to force the output of the forward net to match the desired output. Use backpropagation, but hold the forward net fixed. Only the inverse net is adjusted.*

5. *Series learning and parallel learning are alternated until convergence of the error signals is obtained at both stages.*

6. *After learning is completed, recover the input for the next trial, u_{k+1} from the output of the inverse net.*

7. *Set $k = k+1$ and GOTO Step 2 (Stage 1).*

A key aspect of this algorithm is the alternation between Stage 2 and Stage 3 during learning. This ensures that we have adjusted the net at each trial to invert our best approximation to the plant (in the direction of the desired output), while preserving the integrity of the inverse model for the actual input/output data.

To illustrate the technique, consider the plant mentioned above, from Narendra and Parthadarathy's paper.

$$y_{k+1} = \frac{y_k y_{k-1}(y_k + 2.5)}{1 + y_k^2 + y_{k-1}^2} + u_k.$$

The performance of the learning control scheme when applied to this plant is shown in Figure 7.7 (the desired signal is a triangle). The solid line in the top graph is the desired system output (a triangle). The dotted line is the actual output. The bottom graph shows the corresponding control signal $u_k(t)$. As can be seen, within about ten trials the actual performance closely matches the desired performance. For this simulation each net had two hidden layers, with eight neurons in each hidden layer, three input nodes, and a single output nodes.

7.4 REINFORCEMENT LEARNING CONTROLLER USING AN ANN

The previous two sections presented methods for iterative learning control that used feedforward neural nets with backpropagation training. These schemes, when

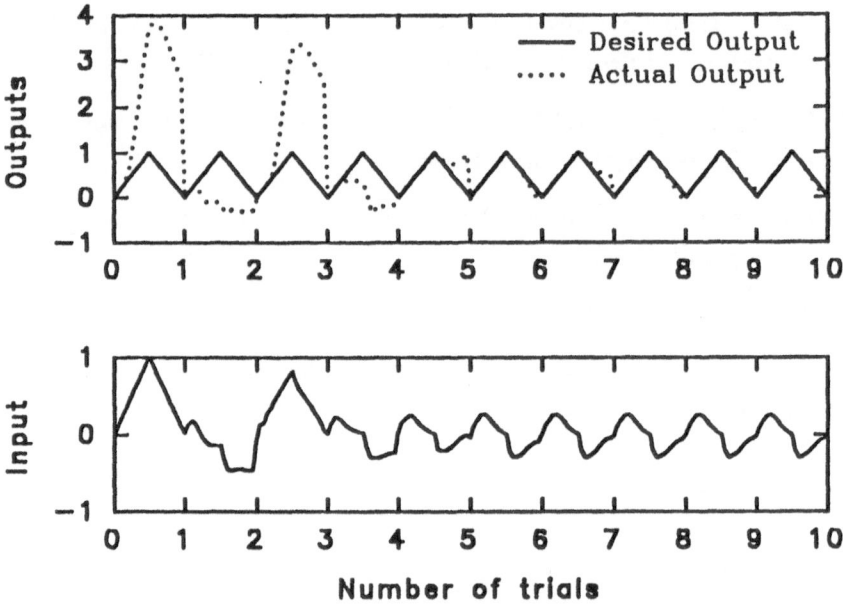

Figure 7.7: Dynamical learning controller performance.

applied to a deterministic plant always result in a completely deterministic system. In this final section we consider a different type of learning controller [1]. This system uses a learning algorithm that is based on reinforcement learning. An important feature of this method is that the learning involves a stochastic search for the best control policy. As a result, even though we apply the method to a deterministic system, the closed-loop system becomes stochastic.

[1]The results reported here were motivated by research sponsored by the Air Force Office of Scientific Research/AFSC, United States Air Force, under Contract F49620-88-C-0053. The United States Government is authorized to reproduce and distribute reprints of this section for governmental purposes notwithstanding any copyright notation hereon.

7.4.1   Reinforcement Learning

An overview of the basic idea of reinforcement learning in neural nets is given in Appendix B. As noted there, a research group based around Barto, Sutton, and Anderson has been very active in this area. An early application of reinforcement learning by their group was to the classic cart-pole (inverted pendulum) problem [78]. A key feature of this work was that it used an input decoder to quantize or partition the state space into "boxes". They used a network made up of what they called "neuron-like" elements, which were basically perceptrons with stochastic outputs. The network applied control forces to a pole-cart system in an attempt to balance the pole. Learning (adjustment of the weights, based on a reinforcement signal from the system) took place only when the pole fell over. This provided negative reinforcement to the system. Using this method, the system was able to learn which actions would keep the pole balanced. They also introduced the idea of trying to predict the value of the reinforcement signal to allow faster learning. One limitation in the work of Barto, Sutton, and Anderson was that it allowed only binary control outputs. Gullapalli [79] was able to extend their ideas to allow real-valued outputs and Franklin [80] has applied this idea to the control of a simulated robot arm. Anderson [81] has also considered the use of multilayer networks with reinforcement learning. Anderson's work also used binary outputs, but the input decoder was replaced by a neural net that learns to distinguish the system state without quantizing it into boxes. Also, his learning algorithms are based on the theory of temporal differences, which has been introduced by Sutton [82], and include a combination of reinforcement learning theory and gradient descent rules.

7.4.2   Proposed Learning Control System

Figure 7.8 shows a proposed configuration we have considered for the iterative learning control of a nonlinear system. This scheme was derived from earlier ideas the author developed on stochastic learning for identification and control in standard control systems [83]. The method combines ideas from both Franklin and Anderson by using a multilayer neural net and real-valued control outputs. Specifically, there are two parts to the learning controller: an action part and a predictor. The control output u is a normal random variable, with mean equal to the output of the action net μ, and with standard deviation given by the predicted reinforcement σ. The idea is that if the predictor expects the proposed action μ

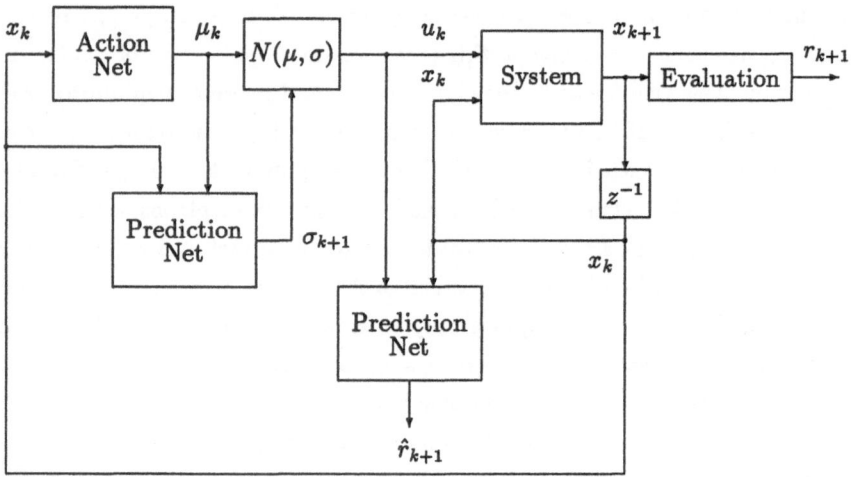

Figure 7.8: Reinforcement learning controller.

to result in a large reinforcement signal (i.e., negative or bad reinforcement), then the actual control action is computed as a random variable centered at μ, but with a large variance. Thus the system is less likely to take an action close to μ. On the other hand, if σ is small this means the predictor expects a more favorable reinforcement, and thus allows the control action to be closer to μ.

Both the action net and the predictor net outputs are defined by equations of the form

$$
\begin{aligned}
z &= \sum_i c_i w_i + \sum_j b_j v_j, \\
v_j &= f(\sum_i a_{ji} w_i),
\end{aligned}
$$

where z is the net's output, the v_j's are the outputs of the hidden layer, the w_i's are the inputs to the net, and $f(\cdot)$ is a sigmoidal nonlinearity. Note that in our scheme, the action net uses only the system states x_k as inputs. The predictor net receives as inputs the system states and either the control action u or the expected control action μ, depending on whether the net is being used in a learning mode or in an operational mode, respectively.

94

The weights of the predictor net are continuously adjusted using the error between the output $\hat{r}_{k+1}$ and the actual reinforcement r_{k+1}. Thus, the predictor learns the reinforcement resulting from a given action u_k, within a given context x_k. Training of this net is done using standard backpropagation with an error signal defined as $e = r - \hat{r}$. This gives

$$b_{i_{new}} = b_{i_{old}} + \eta_b e v_i,$$

$$c_{i_{new}} = c_{i_{old}} + \eta_c e w_i,$$

$$a_{ij_{new}} = a_{ij_{old}} + \eta_a e[b_i v_i(1 - v_i)]w_j.$$

Once the weights of the predictor are adjusted on a given trial, it is then used in an operational mode to compute the variance σ_{k+1} needed to compute the next control action. Note that after learning we have $\sigma_{k+1} = \hat{r}_{k+1} = r_{k+1}$.

Learning in the action layer is a combination of conventional backpropagation and reinforcement learning. The same type of backpropagation equations given for the predictor are used, except that the error signal is modified to be

$$e = R_p \cdot (u - \mu),$$

where

$$R_p = \left\{ \begin{array}{ll} 1 & \dots \text{ if } \sigma_{k+1} > r_{k+1} \\ 0 & \dots \text{ otherwise.} \end{array} \right.$$

R_p is a flag that indicates a desirable system response to the most recent action. Specifically, this signal provides a mechanism for positive reinforcement. The idea behind this algorithm is the following. Suppose that a given control action u results in an actual reinforcement that is smaller than the predicted reinforcement. Then we should adjust the weights so that for the same set of inputs the mean signal μ is moved closer to the actual control action u that gave a better reinforcement. However, if the reinforcement is larger than expected, then we make no changes, but simply allow the stochastic search for a better action to proceed on the next trial.

7.4.3 Example and Comments

This scheme was applied to the same example plant that we used in the previous section,

$$y_{k+1} = \frac{y_k y_{k-1}(y_k + 2.5)}{1 + y_k^2 + y_{k-1}^2} + u_k.$$

In our example, the goal was to train the system output to follow one cycle of a square wave input. In order to do this the configuration of Figure 7.8 was modified to allow input of the desired reference signal to each net. The reinforcement signal for the example was taken as the error between the system output and the reference signal. The predictor provided an estimate of this error, and training of the action net occurred whenever the actual error was smaller than the predicted error. In general the scheme gave good results. Although the net architecture is somewhat different, the predictor could identify the system dynamics with the same results as the identifier used in Narendra and Parthasarathy's paper. We also found the control action to be successful. Figure 7.9 shows a sample result after about 20,000 time steps, for the case of a square wave reference signal. In this trial both the predictor net and the action net had five hidden layer nodes and all the learning gains were set to be 0.01. Note that the square wave was very hard to track. However, in various simulations that we have run, we have found that sinusoids (or sums of sinusoids) can be tracked with essentially no error using our reinforcement learning method.

We feel there are several aspects about reinforcement learning schemes that make them potentially useful for the control of the nonlinear systems. In particular, because the training is stochastic in nature, the neural net will have learned to respond to a variety of input conditions. Thus we expect it to be fairly robust with respect to parameter variations or unmodelled dynamics. We are encouraged by our preliminary results, but more work remains to be done. We are particularly interested in how these ideas could be applied to response shaping and we are currently working to develop a system that learns in response to a time-varying reinforcement signal. Such a signal could be bounded above and gradually grow to zero to reflect overshoot and settling time requirements, respectively. We are also in the process of performing a detailed analysis of the behavior of the proposed control scheme, including additional simulations and the application of the scheme to more realistic systems, such as a six degree-of-freedom aerospace vehicle or a robotic manipulator.

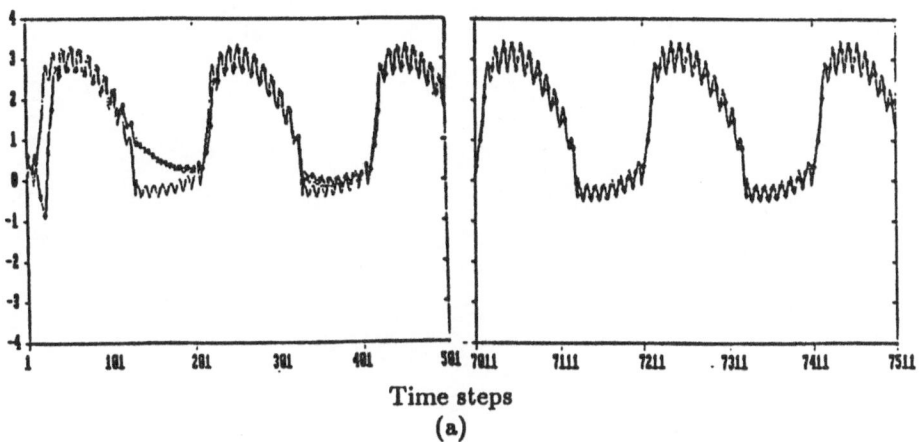

Time steps

(a)

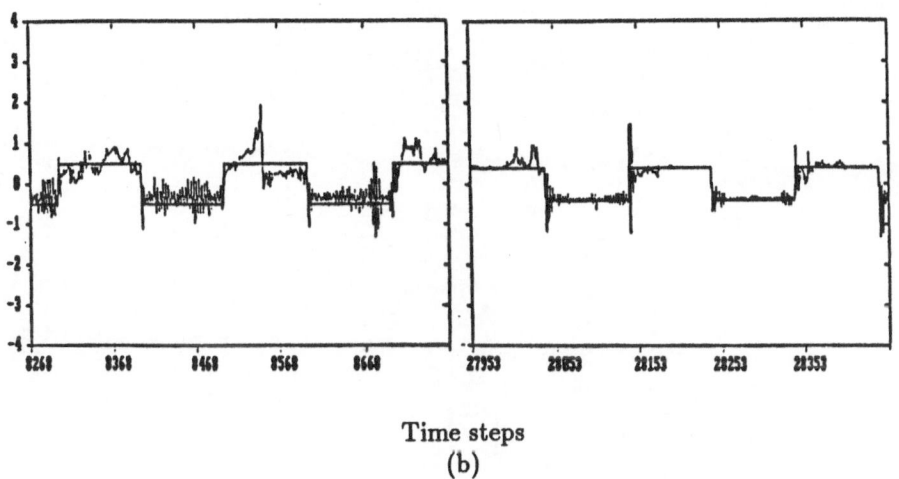

Time steps

(b)

Figure 7.9: Reinforcement learning controller performance: (a) identification; (b) tracking response.

97

CHAPTER 8
CONCLUSION

8.1 SUMMARY

In this research monograph we have presented a number of results related to the analysis and design of iterative learning control systems, motivated by the problem of transient response control in deterministic systems. The monograph began with a description of the concept of learning control and its motivation, a summary of the status of the problem, and a general problem formulation.

We then presented a unifying analysis of the learning control problem for linear time-invariant (LTI) systems. In this analysis we showed that the essential effect of a properly designed learning control scheme is to iteratively produce the output of the best possible inverse of the system. We also showed that if the plant is known, then the best possible learning controller is no better than the best possible controller designed *a priori* from knowledge of the plant. In either case, we obtain an open-loop control law that passes the desired output through the best approximation to an inverse system for the plant. This implies that when the plant is known there is no need in applying an iterative learning control scheme. Further, when the plant is unknown, we showed that it is possible to devise a learning control scheme, based on parameter estimation, that converges with a minimum normed error. However, we noted that the same results can be obtained using a conventional adaptive control scheme. We also showed that when we consider the finite-horizon problem, it is possible to demonstrate learning control with memory, so that the parameters of the inverse system are derived in the learning process. However, for the LTI case these parameters can also be obtained after a single trial, even if the plant is unknown. Again, this implies that for LTI systems learning control offers no fundamental advantage over conventional techniques. This analysis offers insight into the nature of iterative learning control and highlights its capabilities and limitations.

The analysis of LTI learning control suggested that the real usefulness of the technique is for the control of nonlinear or time-varying systems. To illustrate this, we discussed the issues of nonlinear learning control. Then we presented a time-varying learning control technique. This technique can be applied to the class of nonlinear systems that includes typical models of n-jointed manipulators. The approach utilized an adaptive gain adjustment technique, together with the learning control scheme of [7], to iteratively improve the performance of a robotic manipulator. Simulation results verified the effectiveness of the gain adjustment procedure.

Finally, we considered the application of artificial neural networks to the learning control problem. We presented a discussion of the use of neural networks in control systems and then gave three different ways of using an artificial neural network structure for the learning controller. In all three cases, we were able to exploit the ability of a neural network to implement and learn nonlinear mappings from input to output space. Examples were given to illustrate these techniques.

8.2 DIRECTIONS FOR FUTURE RESEARCH

We believe that the results presented in this research monograph are useful. We have established some conclusions regarding the capabilities and limitations of iterative learning control for LTI systems. We have also demonstrated the potential of the technique for nonlinear systems. As is often the case, however, this research has uncovered a number of interesting questions for future study.

Although we feel that the significant analysis is complete for LTI systems, there are a number of extensions that could be considered. For instance, we might consider learning control of LTI systems for enlarged classes of learning controllers, such as time-varying or non-causal learning controllers. Additionally, more work must be done on the issues of convergence rate and robustness. Related to the issue of robustness is the question of LTI learning control when the plant is completely unknown (structure and parameters).

For nonlinear systems, a unifying theory of the analysis and design of iterative learning control systems is not yet available, but is highly desirable. Learning control has the potential to be very useful in robotics and automation, but more basic research is needed. Of particular importance in both LTI and nonlinear learning control is the question of control input magnitude constraints. This must be addressed before significant applications can be realized, because realistic systems will always have limits on the maximum signals attainable from their actuators. Finally, there is a great deal of research needed in the area of neural networks for learning control before significant applications can be achieved.

APPENDIX A

SOME BASIC RESULTS ON MULTIRATE SAMPLING

In this appendix we give some basic results on multirate sampling in sampled-data systems. These results were used in developing the multirate learning control scheme presented in Chapter 5. For more details, including proofs of the results described, the interested reader can refer to [6] or [63]. We begin with an introduction to multirate sampling. Then we describe the modelling of multirate sampled-data systems. Finally, we summarize the main results that are important for the purposes of this monograph.

A.1 INTRODUCTION

A fundamental obstacle in the design of control systems for specified transient response is the presence of the plant zeros in the closed-loop transfer function. If such zeros are in the left-half plane (LHP), it may be possible to design the controller to cancel them [84]. However, when the plant is non-minimum phase (i.e., has zeros in the right-half plane (RHP)), the resulting closed-loop system will always contain the same zeros as the plant (including the ones in the RHP) [85]. Thus it may be impossible to obtain a desired closed-loop transfer function using conventional methods. Recently, multirate digital control techniques have been suggested for overcoming the problems associated with non-minimum phase systems.

In a multirate digital control system, the input and output signals of the plant are sampled at different rates. Such a scheme typically comes in one of two flavors. In the first approach, the input is changed N times more often than the output is sampled. We will refer to this as N-delay input control. This is also referred to as periodic control because the control signal can be viewed as alternating periodically between N different control inputs. In this sense, multirate control can be thought of as a special type of time-varying compensation. The second approach is to sample the output of the continuous plant N times more often than the input to the plant is changed. We call this N-delay output control. Figure A.1 illustrates the difference between N-delay input and N-delay output digital control. In this figure, ZOH stands for zero-order hold and the notation T/N denotes the idea that N samples are taken (or the input is changed N times) in the interval $[kT, (k+1)T]$. We assume the samples are separated by uniform intervals, but in general such an assumption is not necessary. For multivariable systems we may also consider schemes that utilize different numbers of output samples and input changes for each channel.

The properties of multirate digital control systems have been studied by a number of researchers. For an overview of this area see [86] and [87]. Briefly, the attraction of multirate control is found in the additional design freedom that can be attained over conventional single-rate digital control. In particular, it is now well-established that (i) arbitrary (symmetric) pole placement in the closed-loop system can be obtained in both N-delay input and N-delay output schemes and (ii) using N-delay input control allows arbitrary placement of both the poles and the zeros of the discrete-time closed-loop transfer function that relates the input to the output at the sample points.

A.2 N-DELAY DISCRETIZATION

Consider the N-delay input system shown in Figure A.1. The continuous-time plant is described by

$$
\begin{aligned}
\dot{x} &= A_c x + B_c \bar{u}, \\
y &= C x,
\end{aligned}
$$

where $A_c \in R^{n \times n}, B_c \in R^{n \times q}, C \in R^{q \times n}$. The input $\bar{u}$ is generated as the output of the sample-and-hold operation shown in Figure A.1. Assume that

$$
0 < m_2 < \cdots < m_N < 1,
$$

and for convenience define $m_1 = 0$ and $m_{N+1} = 1$. Also define

$$
\begin{aligned}
x(k) &= x(kT), \\
y(k) &= y(kT), \\
u_i(k) &= \bar{u}((k + m_i)T) = u_i(kT),
\end{aligned}
$$

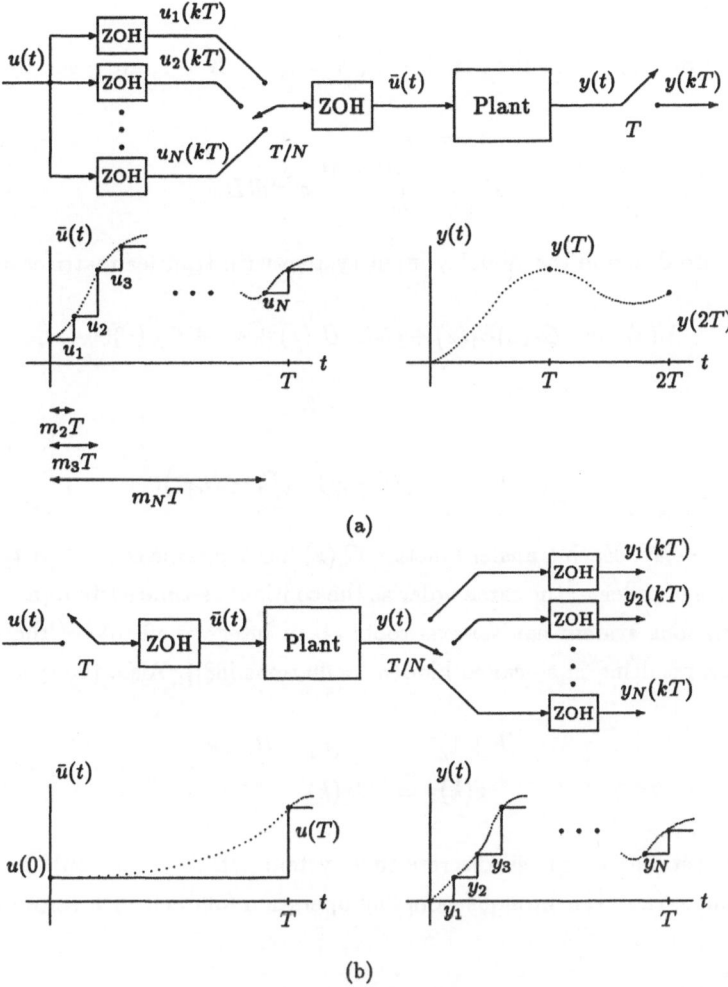

Figure A.1: N-delay control: (a) N-delay input control; (b) N-delay output control.

for $i = 1, \cdots, N$. The resulting sampled-data system can then be modelled by the following discrete-time system (see [63] for details of the derivation).

$$
\begin{aligned}
x(k+1) &= Ax(k) + (B_1 - B_2)u_1(k) + \cdots + (B_{N-1} - B_N)u_{N-1}(k) + B_N u_N(k), \\
y(k) &= Cx(k),
\end{aligned}
$$

where the matrix A and the matrices B_i, $i = 1, \cdots, N+1$, are given by

$$
\begin{aligned}
A &= e^{A_c T}, \\
B_i &= \int_0^{(1-m_i)T} e^{A_c t} dt \, B_c.
\end{aligned}
$$

We can also describe the N-delay input system with transfer matrices as

$$
Y(z) = G_1(z)U_1(z) + G_2(z)U_2(z) + \cdots + G_N(z)U_N(z),
$$

where

$$
G_i(z) = C(zI - A)^{-1}(B_i - B_{i+1}),
$$

for $i = 1, \cdots, N$. Each transfer function $G_i(z)$ has the same characteristic polynomial and is of order n, the same order as the continuous-time system (except when the continuous system has $j\omega$-axis roots at an integer multiple of the sampling frequency, resulting in so-called hidden oscillations [88]). Also, the system

$$
\begin{aligned}
x(k+1) &= Ax(k) + B_1 u(k), \\
y(k) &= Cx(k),
\end{aligned}
$$

defines the standard or basic discrete-time system (BDTS) that would be obtained with a single-rate, synchronous sampling operation followed by a zero-order hold. Thus,

$$
G(z) = C(zI - A)^{-1}B_1
$$

is the usual impulse invariant z-transform model of the plant preceded by a zero-order hold. Another interesting fact is that

$$
G(z) = G_1(z) + G_2(z) + \cdots + G_N(z).
$$

Analogous expressions can be developed to describe the intersample behavior of the multirate sampled system [63]. These are generalizations of the standard modified z-transform, which is commonly used to describe nonsynchronous sampling operations and intersample behavior [89].

With the modelling of the sampling as described, we can view the N-delay sampling schemes as follows. When the input is changed more often than the output is sampled, the effect is to create an N-input, single-output system with respect to the output samples $y(kT)$. This is shown in Figure A.2, where the transfer functions $G_i(z)$ are defined as above. In the same way, if we sample the

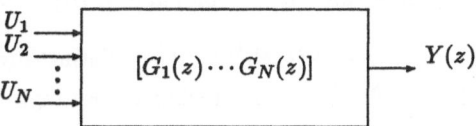

Figure A.2: N-delay input control: an MISO system.

output of a continuous system at N points (say $y(kT + n_iT)$, $i = 1, \cdots, N$), we effectively create a single-input, N-output system with respect to the N sequences of output samples. This is shown in Figure A.3 (here the $G(z, n_i)$ generalize the modified z-transform).

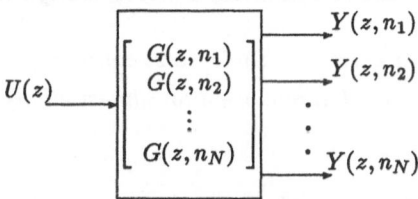

Figure A.3: N-delay output control: an SIMO system.

107

A.3 BASIC RESULT

The result that is useful for the multirate iterative learning controller of Chapter 5 was given by Mita and Chida [5]. For completeness, however, we give an obvious generalization of their result. Then we show how a multirate sampling scheme increases the design freedom available in a control system by considering the use of N-delay input and N-delay output multirate sampling in a closed-loop configuration. The following assumptions on the open-loop system are needed.

(A1) The continuous-time system (A_c, B_c, C) is controllable and observable.

(A2) The sampling period T is such that $G(z)$ has no zero at the origin, the BDTS (A, B_1, C) is controllable and observable, and $|CB_1| \neq 0$.

(A3) The eigenvalues of $\alpha = A - B_1(CB_1)^{-1}CA$ are distinct or, if multiple, belong to a single Jordan block. (This is a technical assumption, which is always satisfied for single-input, single-output systems that satisfy assumption A1 [5]. Note that the $n - q$ nonzero eigenvalues of α are the zeros of $G(z)$.)

Now, the zeros of the N-delay input system are defined to be those values of z for which the system matrix $R(z)$ loses rank below $n + q$, where $R(z)$ is given by [90]

$$R(z) = \begin{bmatrix} A - zI & (B_1 - B_2) & (B_2 - B_3) & \cdots & (B_{N-1} - B_N) & B_N \\ C & 0 & 0 & \cdots & 0 & 0 \end{bmatrix}.$$

Mita and Chida proved the following theorem for the special case of $N = 2$.

Theorem A.1 *([5], Theorem 2): Under assumptions (A1)–(A3), with $N = 2$, the system matrix $R(z)$ is of full rank $n+q$ for almost all m_2 satisfying $0 < m_2 < 1$ and for arbitrary z.*

We comment that "almost all" means that given a fixed m_2 it is possible to construct a system (A_c, B_c, C) such that the rank condition fails. However, given a system (A_c, B_c, C), Theorem A.1 holds everywhere except on a set of measure zero. In [63] we proved the following extension of this theorem.

Corollary A.1 *Under assumptions (A1)–(A3), for any N, the system matrix $R(z)$ is of full rank $n + q$ for almost all $0 < m_N < 1$ and for arbitrary z.*

For a single-input, single-output plant (i.e., $q = 1$), this result means there is no zero common to all the transfer functions $G_i(z)$, for $i = 1, \cdots, N$. This is the key property that allows us to place the closed-loop system zeros in a multirate control scheme and is the property that we use for the multirate iterative learning controller. A dual result can be given for the case of N-delay output control.

To illustrate the usefulness of these results, consider the closed-loop configurations shown in Figure A.4. In Figure A.4(a) we note that, because the N-delay input control effectively makes the system an MISO scheme, it is possible to use a separate feedforward compensator for each input channel. For this configuration, the closed-loop transfer function from $R(z)$ to $Y(z)$ is

$$\frac{Y(z)}{R(z)} = \frac{G_1 C_{11} + \cdots + G_N C_{N1}}{1 + G_1 C_{12} + \cdots + G_N C_{N2}}.$$

In this transfer function the denominator is affected by the C_{i2} compensators and the numerator is controlled by the C_{i1}'s. Because the G_i's have no common zeros, the numerator and denominator of the closed-loop system can be assigned independently. Also, as we increase N we can decrease the order of the individual C_{ij}'s.

Similarly, for the N-delay output scheme of Figure A.4(b) the resulting closed-loop system is

$$\frac{Y(z)}{R(z)} = \frac{C_{FF} G(z)}{1 + G(z, n_1) C_1 + \cdots + G(z, n_N) C_N}.$$

For this case we see that we have increased the number of free control parameters in the denominator over a single-rate digital control scheme. However, the numerator cannot be changed beyond the multiplicative feedforward term C_{FF}. Thus the plant zeros will still occur in the closed-loop transfer function. For this reason N-delay input control has more potential for use in shaping the transient response.

One caveat regarding these results is that they apply with respect to the output sample points. Unfortunately, in between the samples the output response may not behave as expected. This is a consequence of the multirate sampling technique and represents a limitation of control schemes that use multirate sampling. More details about this effect can be found in [6] or [63].

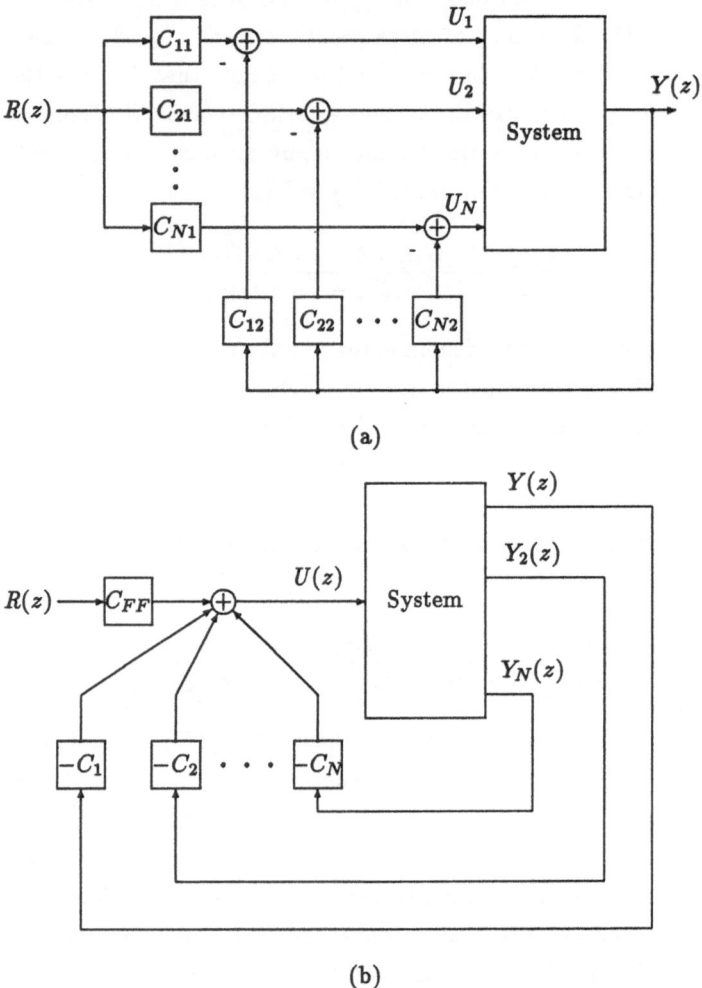

(a)

(b)

Figure A.4: Closed-loop configurations: (a) N-delay input control; (b) N-delay output control.

110

APPENDIX B

TUTORIAL ON ARTIFICIAL NEURAL NETWORKS

So, as fast as I could, I went after my net.
And I said, "With my net I can get them I bet.
I bet, with my net; I can get those Things yet!"

— Dr. Suess, from *The Cat in the Hat.*

The field of artificial neural networks has recently attracted considerable attention from the engineering community. Significant applications of neural nets are emerging in almost every field of endeavor. The purpose of this appendix is to present a self-contained tutorial on artificial neural networks (ANNs for short). The tutorial will provide sufficient background for the reader to understand the neural net-based iterative learning control schemes given in Chapter 7. Additionally, it can be used by those who simply wish to know more about this exciting research field. We will begin with a description of ANNs. We also give a brief summary of the history of the field. Then we look at the properties of ANNs and describe some applications. For completeness, we also include a derivation of the backpropagation learning algorithm that is often used for training multilayer feedforward neural networks. Finally, we point out some essential references that the interested reader may consult for more detailed information on artificial neural nets. Throughout the appendix we will interchangeably use the expressions artificial neural network, neural net, ANNs, or simply nets. We will also adopt a more informal style in this appendix than we have used in the body of the monograph[1].

---

[1]The style reflects the fact that this appendix is adapted from a tutorial article published by the author in the February 1992 issue of the student magazine, *IEEE Potentials* (vol. 11, no. 1, pp. 23–28).

B.1 AN INTRODUCTION TO NEURAL NETWORKS

An ANN is an artificial (man-made) system motivated by the neural structure observed in real biological organisms. The phrases "neural net" or "neural network" actually came from the literature of neurophysiology, the study of how the brain and its central nervous system work. A biological neural network is a collection of a large number of neurons connected together in various ways (neurons are the cells that are the basic building blocks of the nervous system). Neurophysiologists studying the cerebral cortex and other parts of the nervous system developed mathematical models of these biological neural nets to describe their operation. As people began to analyze these models, they found that they exhibited a number of useful and interesting properties – properties similar to those of the biological systems they described, such as adaptability, learning, feature classification, and generalization of learning from past experiences to new experiences. Soon researchers began to study ways of building electronic networks that implemented the same math models, using the structures observed in the biological systems. Hence the term "artificial" neural networks.

Actually, the attempt to build artificial systems based on the structure we have observed in biological systems is quite natural. For instance, when we consider the motor behavior of living organisms we encounter the perfect control system. Biological systems possess the ability to perform complex movements with precision and grace, often in the presence of disturbances. They can adapt to changes in their environment and can learn to recognize and respond to stimuli. They can also generalize from past experience. A theory for the design of systems that even partially achieve the performance of biological systems would be a welcome contribution in many fields, including computer science, robotics, automation, and others.

(As an aside, we would mention that the protagonists in the various books of Isaac Asimov's classic robot stories, from *I, Robot* to *Robots and Empire*, were a marvel of science and technology. The robots of Asimov's imagination were capable of both human-like movement and human-like thinking, reasoning, and decision-making abilities. To describe the science of building robots with "artificial intelligence" he coined the term positronics. A positronic brain was not simply a big, powerful computer with a long computer program. Instead, it was a combination of complex software coupled with a dense, highly interconnected electronic network, organized to produce collective behavior through its connections.

In his vision of the future, Asimov's ideas were surprisingly close to reality. With the discovery of math models with the ability to reproduce some features of the brain, together with recent progress in VLSI technology that makes it possible to build extremely dense electronic circuitry, many believe that the positronic brain of Isaac Asimov's science fiction may someday be plausible.)

An artificial neural network can be formally defined by three elements: (1) a set of processing elements, called neurons; (2) a specific topology of weighted interconnections between these elements; and (3) a learning law that provides for updating the connection weights. We describe each of these elements separately.

B.1.1 Neurons

A typical neuron is shown in Figure B.1. In this type of neuron, called a perceptron, the output is a nonlinear function of a weighted sum of its inputs. The input/output relation is often described as the neuron's transfer function, although the relationship is usually specified in the time domain. One of the simplest (and most common) transfer function is defined by

$$y = f(\sum_{i=1}^{n} w_i x_i - \theta),$$

where the w_i's are the weights, θ is a threshold, and $f(\cdot)$ is a nonlinear function. Some nonlinear functions that are often used are also shown in Figure B.1. Each of these types of nonlinearity is motivated by observations of biological neurons. Note that a neuron may have a simple algebraic transfer function as we have described, but it could also include dynamic behavior, such as integration or feedback of its own output as one of its inputs. For example, such a neuron might be described by an equation of the form

$$\dot{y} = a(y) + f(\sum_{i=1}^{N} w_i x_i - \theta),$$

where $\dot{y}$ denotes the derivative of y with respect to time and $a(\cdot)$ is a specified function.

B.1.2 Interconnection Topology

The transfer function of the neuron is one way to distinguish neural networks. A second way is by the topology of interconnection of the neurons to one another.

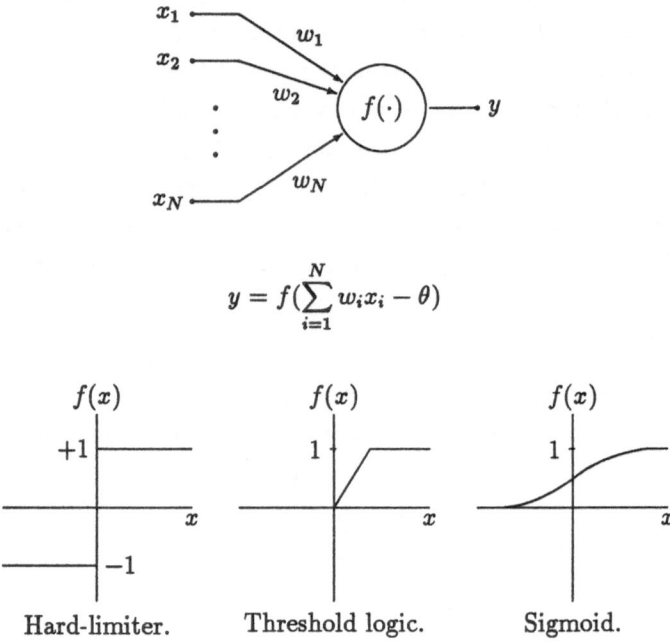

$$y = f(\sum_{i=1}^{N} w_i x_i - \theta)$$

| $f(x)$ | $f(x)$ | $f(x)$ |
| Hard-limiter. | Threshold logic. | Sigmoid. |

Figure B.1: Typical neuron.

Neural network topologies fit broadly into two classes: feedforward and recursive. A typical feedforward net is shown in Figure B.2. This network has an input layer (shown simply as a fan-out from the inputs), two "hidden" layers, and an output layer. Each output in a layer is connected to each input in the next layer. A typical set of equations for such an architecture might be the following:

$$y_k = \sum_{i=1}^{M} w_{ki} v_i,$$

$$v_i = f(\sum_{j=1}^{M} \alpha_{ij} z_j),$$

$$z_j = f(\sum_{l=1}^{N} \beta_{jl} u_l),$$

where

$$f(x) = \frac{1}{1 + e^{-ax}}.$$

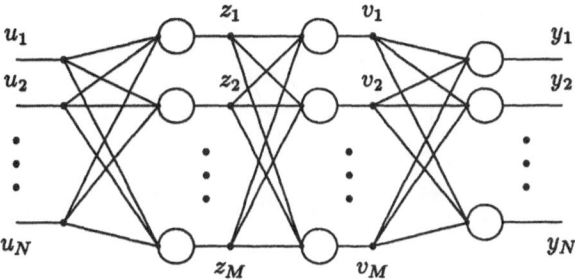

Figure B.2: Feedforward neural network with two hidden layers.

In this case we have indicated that the output layer has simple linear neurons, while we have specified that all the neurons in the two hidden layers have the same transfer function, with a sigmoidal nonlinearity. However, these assumptions are not necessary. In general, the number of hidden layers in the net and the number of neurons in each hidden layer may be varied, along with the transfer functions of the individual neurons, the number of outputs, and the number of inputs. Also notice that, because there is no feedback between layers, the effect of the feedforward neural net topology is to produce a nonlinear mapping between the input nodes and the output nodes. As long as the interconnection weights are fixed, this mapping is completely determined.

The other broad class of ANNs is the class of nets that have feedback. Such nets are called recursive. In a recursive ANN, each neuron receives as its input a weighted output from every other neuron in the net, possibly including itself. An example of a recursive network with N neurons is the Hopfield network with asynchronous updating (due to John Hopfield). Such a net is shown in Figure B.3. This system receives an input pattern of u_i's and then iterates according to the equation shown in the figure. The nonlinearity can be a hard limiter or a threshold logic function. A key feature of Hopfield's original network was that the iteration is done randomly. That is, a neuron is chosen at random and then its output is changed according to its transfer function. This process continues until the outputs (the x_i's) reach a steady-state equilibrium point. in [91], Hopfield showed that convergence to a stable equilibrium point is assured if the weight matrix $W = [w_{ij}]$ is symmetric (i.e., $w_{ij} = w_{ji}$) and the diagonal elements w_{ii} are

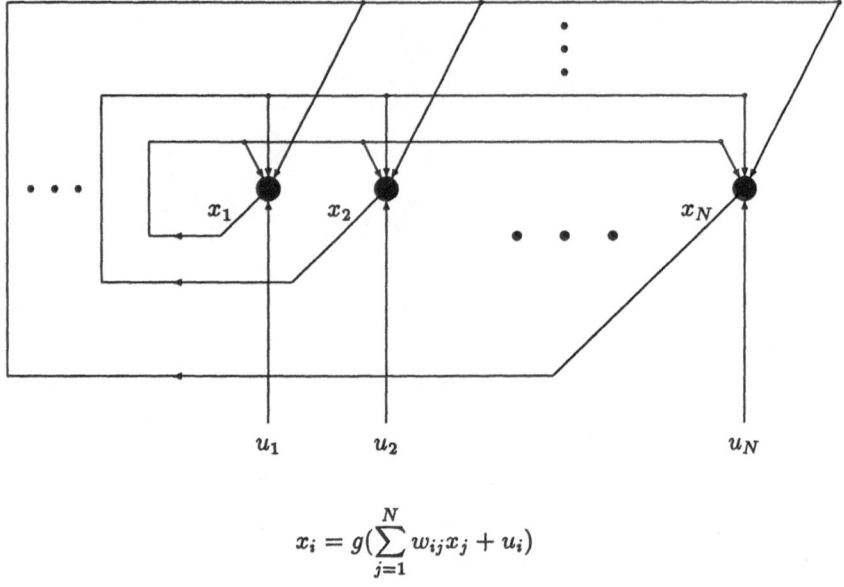

$$x_i = g(\sum_{j=1}^{N} w_{ij}x_j + u_i)$$

Figure B.3: Recursive neural network.

zero.

Other types of recursive ANNs have been studied by a number of researchers. One important class is the continuous-time Hopfield model [92], described by N equations of the form

$$C_i \frac{dx_i}{dt} = -\frac{x_i}{R_i} + \sum_{j \neq i} w_{ij}g(x_j) + u_i.$$

Another type of recursive neural net is defined by N nonlinear differential equations of the form

$$\dot{x}_i = a_i(x_i)[b_i(x_i) - \sum_{j=1}^{N} d_k(y_k)].$$

This is a quite general model, which has been motivated by biological considerations. It includes as special cases many ANNs found in the literature, depending on the functions $a(\cdot), b(\cdot)$, and $d(\cdot)$. This class of neural net models has been extensively studied by Stephen Grossberg, one of the most prolific researchers in the field [93].

Notice that recursive neural nets have a more complex behavior than feedforward nets. In a feedforward net an output appears almost immediately following the application of an input vector. The only delay is due to the propagation through the hidden layers. On the other hand, in a recursive net the outputs of the system evolve with time until a steady-state equilibrium point is reached. Thus, the output of a neural network with a recursive topology is defined by a trajectory (or sequence) of vectors, rather than by a single vector.

B.1.3 Learning Laws

The third distinguishing characteristic of a neural network is the learning law that is used to change the interconnection weights. To understand this, note that ANNs are usually used in one of two ways. In the first way the neural network is simply used to compute an output in response to a given input. This is called an operational or computation mode, and assumes that the interconnection weights have been fixed to some desired values. The second way a neural network may be used is in a learning mode. In a learning mode the object is to adjust the weights so that the neural network output has some desired property. Of course, in most applications both modes are present. The ANN is first used in a learning mode, with training proceeding until the weights are correctly adjusted for the particular application. Then the net is used in the operational mode.

Learning algorithms for weight adjustment can be described as either supervised learning, unsupervised learning, or reinforcement learning techniques. Supervised learning, also called learning with a teacher, assumes that the desired output of the neuron is known. The error between the desired output and the actual output is used to form an error signal, which is used to update the weights according to a prescribed learning algorithm. In unsupervised learning the desired output is not known, but instead learning is based simply on input/output values. Such learning algorithms usually act to extract features from sets of input data. Finally, in a reinforcement learning technique, weights associated with a neuron are not changed proportional to the output error of that particular neuron, but instead are changed in proportion to some type of global reinforcement signal. Such a signal may give qualitative measures of performance, such as good or bad (e.g., $+1$ or -1). Examples of each of these types of learning techniques will be given below.

B.2 HISTORICAL BACKGROUND

Although the field of artificial neural networks has experienced a kind of second birth in recent years, it actually has such a rich history that it would take a lengthy survey article to give a complete description. In the next few paragraphs we will try to give a sketchy outline of the past work in the field. This review represents the author's biases and has no presumption of completeness. To learn more the reader is urged to consult the references given in the final section.

The seminal work in the field of ANNs is widely regarded to be that of Warren McCollough and Walter Pitts' paper, "A Logical Calculus for the Ideas Immanent in Nervous Activity," appearing in the *Bulletin of Mathematical Biophysics* in 1943 [94]. The model of a neuron we showed in Figure B.1, with a threshold logic nonlinearity, was introduced in this paper (50 years ago!). An array of such neurons was called an M-P network and the properties of this model were studied by a number of researchers throughout the 1940's. Another key work, combining ideas from psychology and neurophysiology, was Donald Hebb's 1949 book, *The Organization of Behavior* [95]. This work was one of the first to propose the idea of adaptation of the synaptic strengths during learning. (In a biological neural network, neurons are connected by axons, which communicate the output and input signals. Synapses connect the dendrites (biological fan-in connectors) of the neuron to the axons. The "strength" of the synapse is analogous to the "size" of the weight connecting two neurons.) Using this idea, Frank Rosenblatt showed, in a 1958 paper, how an M-P network could be trained to classify sets of patterns by adjusting the weights of the neurons [96]. In addition to proving the Perceptron Convergence Theorem, Rosenblatt also introduced the name "perceptron" for the M-P network.

Through this same period, researchers were actually building the first primitive cybernetic machines. In 1951, Marvin Minsky and his colleagues at MIT demonstrated a neural network made of electronic units, connected by a network of mechanical linkages and relay switches. The machine used reinforcement learning to produce a desired output. Six years later, Rosenblatt demonstrated the Mark I Perceptron. This early neurocomputer, built with vacuum tubes and motor driven potentiometers, functioned as a character recognizer.

In the 1960's the research in ANNs continued. Bernard Widrow introduced the ADALINE, which stands for ADAptive LINear Element, and the MADALINE (a multilayer ADALINE) [97]. The ADALINE had the same transfer function and

118

topology as a single-layer perceptron, but its learning rule was different. By the middle of the decade, Widrow had moved his research into the area of adaptive signal processing. Those familiar with this field will notice the similarities between many adaptive signal processing and parameter estimation algorithms and the learning laws from the neural net literature. It is ironic that people are now applying ANNs to signal processing tasks previously performed using algorithms that originated in the neural network research of the early 1960's!

At the same time that Widrow was studying ADALINES, Minsky was continuing to investigate the capabilities of perceptrons. The culmination of this work was a book by Minsky and Papert called *Perceptrons* [98]. The 1969 publication of this volume marked a pivotal point in the history of the field. While the book presented an elegant mathematical approach to the study of perceptrons, it also pointed out the limitations of single-layer M-P networks. In a classic example, Minsky and Papert showed that it was impossible for a perceptron to learn as simple a classification as the exclusive-OR (XOR) operation. In a strange quirk of events, the research activity related to perceptron-based ANNs virtually ceased, almost overnight. Many people have credited *Perceptrons* with the dubious distinction of having single-handedly stifled the field. However, in a 1988 expanded edition of the book Minsky and Papert object to this view. Rather, they argue, the role of the book was to point out the limitations of the theory, which in turn should stimulate the hunt for a better theory. Whatever the cause, the fact remains that in the 1970's and the early 1980's the research activity related to ANNs was almost nonexistent, with only a few dedicated adherents continuing to work in the area. It was during this period that the artificial intelligence community decided on symbolic-based computing methods, using conventional von Neumann processing, as the paradigm of choice for AI, rather than neural networks.

During these "quiet years", as Robert Hecht-Nielsen refers to them in his recent book, *Neurocomputing* [99], some progress was made. In particular, the research of Stephen Grossberg kept the flame going. Grossberg was interested in more complicated models of biological neural networks, such as the general recursive model we described above. One problem that the field encountered was that it was hard! The math that was needed to analyze ANNs and their learning properties was difficult, and in some cases did not exist. One of Grossberg's important contributions was to develop much of the mathematics necessary to get new results. This took time, but by the mid-1980's, with the help of Hopfield, Hecht-Nielsen, James Anderson, Teuvo Kohonen, and a few others, the field was ready to be revived.

In 1986 McClelland and Rumelhart published a two-volume book titled *Parallel Distributed Processing* [100]. The PDP book is thought by many to be the catalyst that catapulted the ANN field to buzzword status. The thing that got people excited about the book was its introduction of the so-called backpropagation algorithm. In *Perceptrons* Minsky and Papert had indicated the limitations of single-layer perceptrons, but they noted that these limitations might be overcome by multilayer perceptrons if there was a good training algorithm for these types of nets. The backpropagation algorithm of the PDP book was what the field needed. McClelland and Rumelhart showed that by using backpropagation it was possible for a two-layer perceptron to learn the XOR mapping. With backpropagation Minsky and Papert's objections were met and suddenly the "new" field of ANNs appeared.

It is interesting to note that research is a funny thing. The backpropagation algorithm was independently derived by Werbos in 1974 [101] and by Parker in 1982 [102]. Somehow neither time was right to nudge the ANN field back to its feet. Perhaps if any one thing made 1986 the right time, it was probably the availability of easily accessible, powerful computers. The computations needed to experiment with learning algorithms, and the hardware necessary to even consider real-time implementation of ANNs, may have held the field back through the 1970's. However, now it is possible to buy neural net plug-in boards for your PC, and anyone can experiment with learning algorithms without tying up a mainframe for days. Thus, it may be that ANNs provide an example of a field that could not exist without the advanced level of computer technology that we have available today.

As we warned earlier, this has been a short and incomplete history, but hopefully it has given you a flavor of the field. It probably goes without saying that the story is not over. We can argue about whether the field was stifled by one book and revived by another. However, it is clear that the current resurgence in activity and the wave of new results promise a bright future for the field of ANNs.

B.3 PROPERTIES OF NEURAL NETWORKS

Artificial neural networks have been found to have many interesting and attractive properties. Applications of these properties are wide and varied. The quote from Dr. Suess at the beginning of the appendix is a fair description of the state of affairs. Think of almost any "thing" you can imagine and chances are that some-one has attempted to deal with it using a neural net. However, in reviewing the literature, we find that most of these "things" fit into one of four broad classes of applications, depending on which particular neural network property is being ex-ploited. These four classes are: (1) pattern classification and associative memory; (2) self-organization and feature extraction; (3) optimization; and (4) nonlinear mappings. In most of these the desired effect is obtained through learning.

B.3.1 Pattern Classification and Associative Memory

By far, the most popular feature of neural networks has been their ability to learn to distinguish between classes of inputs and to then store and recall these associations. To see how this can be done, first consider a simple perceptron, as shown in Figure B.4. Suppose we have two classes, A and B, separated by the

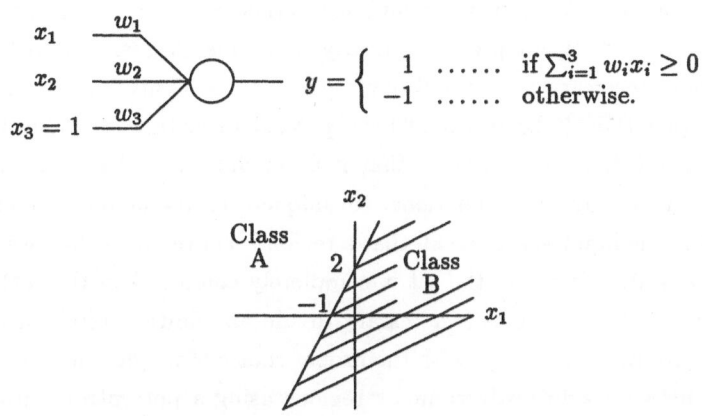

Figure B.4: Pattern classification with a perceptron.

line shown in the figure. Then it is clear that if the weights of the perceptron have the values $w_1 = -2, w_2 = 1$, and $w_3 = -2$, then the perceptron will output a $+1$ whenever the pair (x_1, x_2) is taken from class A, and will output a -1 if the pair is from class B. Thus the net can distinguish between the two classes. Note that in this case we have introduced a threshold by using a constant input $x_3 = 1$. This is often done to expedite programming and also because the threshold may then be adaptively adjusted, just like the weights.

Now, if we knew the line that separated the two classes we could preset the weights to reflect this knowledge. But what if the boundary was unknown? In this case we might be able to adjust the weights to the correct values by presenting pairs from class A and B on alternate trials and adjusting the weights, based on the error in the output. Note that this would be a supervised learning algorithm, because we are assuming that the desired output is known. Specifically, we could use an algorithm known as the Delta rule. Let y_d denote the desired class, $\underline{w}$ denote the weight vector, and $\underline{x}$ denote the input vector. Then the Delta rule says to adjust the weights after each presentation of an input vector, according to

$$\underline{w}_{new} = \underline{w}_{old} + \beta(y_d - y)\underline{x},$$

where β is called the learning gain and is typically less than one. Rosenblatt's Perceptron Convergence Theorem proved that the Delta rule will converge if the classes A and B are linearly separable.

Of course, the XOR function we mentioned earlier is a classic example of a simple function that does not produce linearly separable classes. To deal with this problem it is necessary to use multilayer perceptrons. This is illustrated in Figure B.5. Figure B.5(a) shows a multilayer perceptron with one hidden layer. With this structure, it is easy to show that it is possible, with the right choice of weights and neuron transfer functions, to uniquely represent open or closed convex regions of the input space (recall that a region is convex if the line segment that connects any two points in the set is completely contained in the set). To distinguish non-convex regions, it is necessary to add an additional hidden layer, as shown in Figure B.5(b). In fact, with the proper choice of weights, it is possible to distinguish between arbitrarily complex regions using a perceptron with two hidden layers. As a general rule, the number of neurons in the second hidden layer determines the number of regions we can distinguish, while the number of neurons in the first layer determines the number of edges for each region. An excellent overview of these issues can be found in Richard Lippman's survey paper, "An Introduction to Computing with Neural Nets" [103]. As a final point, notice that we have shown regions with sharp corners in Figure B.5. This is a result of using threshold logic nonlinearities. If we use a sigmoidal function instead, then we can get smooth edges.

The discussion of the previous paragraph was concerned with the ability of a neural network to represent regions of its input space, assuming we knew how those regions were defined. But, how do we train the neural net to recognize these

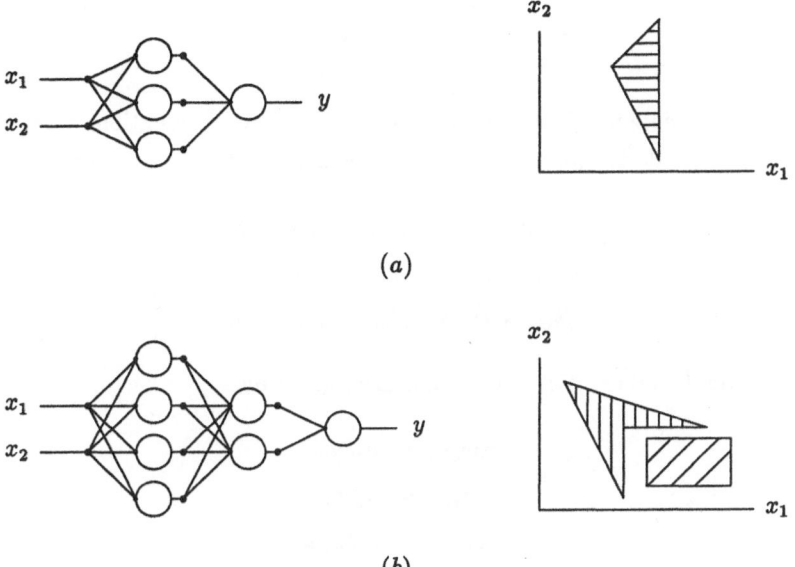

(a)

(b)

Figure B.5: Multilayer perceptrons.

regions when they are not known? This is where the backpropagation algorithm comes in. Also called the generalized Delta rule, backpropagation is an approach to adjusting the weights associated with the hidden layers, based on the error of the output neurons. The idea behind backpropagation is really quite simple. To derive the algorithm, we define an error function for the net, and then compute the partial derivative of this function with respect to the weight we want to adjust. The change in the weight should then be made in the direction of the negative of the partial derivative.

To give a specific example of this, consider again the network of Figure B.2. If we define the error function as

$$E = \frac{1}{2} \sum_{k=1}^{N} (y_k^d - y_k)^2,$$

where y_k^d denotes element k of the desired output vector $\underline{y}_d$, then the backpropagation algorithm says that we should choose weight changes given by

$$\Delta w_{ki} \propto -\frac{\partial E}{\partial w_{ki}},$$

123

$$\Delta\alpha_{ki} \quad \propto \quad -\frac{\partial E}{\partial \alpha_{ki}},$$

$$\Delta\beta_{ki} \quad \propto \quad -\frac{\partial E}{\partial \beta_{ki}}.$$

Assuming sigmoidal nonlinearities, this results in following equations for updating the weights (the derivation of these equations is presented below, in section B.6).

$$w_{new}(k,i) \quad = \quad w_{old}(k,i) + \Delta w_{ki},$$

$$\alpha_{new}(i,j) \quad = \quad \alpha_{old}(i,j) + \Delta\alpha_{ij},$$

$$\beta_{new}(j,l) \quad = \quad \beta_{old}(j,l) + \Delta\beta_{jl},$$

with the update values $\Delta w_{ki}, \Delta\alpha_{ij}$, and $\Delta\beta_{jl}$ given by

$$\Delta w_{ki} \quad = \quad \eta_1 \delta y_k v_i,$$

$$\Delta\alpha_{ij} \quad = \quad \eta_2 \delta v_i z_j,$$

$$\Delta\beta_{jl} \quad = \quad \eta_3 \delta z_j u_l,$$

where we use

$$\delta y_k \quad = \quad y_k^d - y_k = e_k,$$

$$\delta v_i \quad = \quad a v_i (1 - v_i) \sum_{k=1}^{N} w_{ki} \delta y_k,$$

$$\delta z_j \quad = \quad a z_j (1 - z_j) \sum_{i=1}^{M} \alpha_{ij} \delta v_i,$$

for $k = 1, 2, \ldots, N$, $i, j = 1, 2, \ldots, M$, and $l = 1, 2, \ldots, N$. Here the η_i's are proportionality constants, often called the learning gains. The motivation for the name backpropagation can be seen from these equations. Think of the quantities δz_i and δv_j as the "errors" associated with hidden layers one and two, respectively. Then you can see how the "error" at the nodes in the second hidden layer is formed, by taking the actual errors at the output nodes and passing them through the weights connecting the second hidden layer to the output. Thus we have "propagated" the errors from the output "back" through the network weights, to form pseudo-errors for the nodes of the second hidden layer. Similarly, the "errors" of the second layer are propagated back through the next layer of weights, to form "errors" for the first hidden layer neurons.

Backpropagation has been shown to be a successful learning algorithm by a number of researchers. Unfortunately, there is no counterpart to Rosenblatt's

124

Perceptron Convergence Theorem for backpropagation. The reason is that back-propagation simply gives a gradient descent along a cost function in the weight space. As a result, it can get stuck in local minimums. A lot of work has gone into trying to understand why backpropagation works so well in some cases and how it can be improved when it does not converge the way we want. This is still an area of ongoing research.

In addition to being able to distinguish classes of patterns, ANNs can also learn to associate input/output pairs (without any knowledge of the process generating the pairs) and store the associations for later recall. An example of an associative memory is the Hopfield network mentioned above. This network is described by a system of nonlinear differential or difference equations and as such it contains a number of equilibrium points. When run in an operational mode, the network receives an input vector that acts as a forcing function for the system equations. Eventually, the system's trajectory converges to an equilibrium, which defines the output pattern. Because the equilibria of such nonlinear systems usually have a non-trivial basin of attraction, it is possible to converge to the same equilibrium point if we input a vector that is different from the original input, but inside the region of attraction. This feature can be applied to pattern recognition problems in image and signal processing. For instance, suppose we photographed a face with a beard and formed a pixel image, which we stored in an associative memory. By this we mean that whenever the pixel image is applied to the Hopfield net, the system will converge to a given state, which we then associate with the input. If we later photograph the same face after the beard had been shaved, form a new pixel image, and then present this image as an input to the neural net, the net will then converge to the same state, assuming this new pixel image was in the same region of convergence as the original pixel image of the unshaven face. We are leaving out lots of details here, but this is the essential idea behind using recursive nets as associative memories.

Notice that the behavior of a recursive ANN is highly dependent on its weights. In order to use a net as an associative memory, it is necessary to design the weights properly. This involves both the issue of stability (i.e., will the network ever converge to an equilibrium point?) and the question of how to design the weights so that the system has a desired equilibria set. These are hard mathematical problems, but a great deal of progress has been made. As an example, a Hopfield-type associative memory has been reported that can recognize thirty handwritten Chinese characters [104].

B.3.2 Self-Organization and Feature Extraction

A second capability of artificial neural networks is their self-organizing characteristics, studied primarily by Teuvo Kohonen [70,71]. Consider a network with an array of input nodes and an array of output nodes, configured so that every input node is connected to every output node. If you present a series of random input vectors and vary the weights according to an unsupervised learning algorithm, then the weights will adjust so that topologically close output nodes are sensitive to similar inputs. This is called self-organization because there is no teacher during learning and because the neural net has organized itself to extract features from the input data.

It is believed that self-organization is an important feature of biological neural networks, particularly in the sensory-motor and visual systems. For instance, suppose a cross-section of the visual region of your cerebral cortex was taken while you are looking across the room. If you could stain or dye those neurons that were high (i.e., output equal one) then you would find that the neurons would form an outline of the room. If you then shifted your gaze the neurons would reorganize into a new shape to reflect the new sensory input. Another example is the so-called tono-topic organization of the neurons associated with the auditory pathway [70]. It has been found that neurons that are physically close to each other respond to frequencies that are close. For example, if a neuron in the auditory pathway gives the strongest output response to a signal at frequency 440 Hz, then the neuron next to it will be the one that gives the strongest response to a 445 Hz signal. This ability to organize the weights of the net to represent the organization of its inputs can be thought of as a feature extraction property. If a given set of data has an underlying structure or features, the ANN can extract these features through self-organization to form what are called feature maps. These properties have been used in fault detection systems and for automatic target recognition.

To give a very simple example of a self-organizing algorithm, refer to Figure B.6. This net operates a little differently than others we have discussed. In this case the only neuron that outputs a one is the neuron whose weight vector is closest (in the sense of Euclidean distance) to the input vector. All other neurons output a zero. That is, for each neuron define the distance

$$d_i = \sum_{j=1}^{2}(x_j - w_{ij})^2.$$

Then $y_i = 1$ only if $d_i < d_j$ for $i \neq j$; otherwise set the output to 0. Next, define

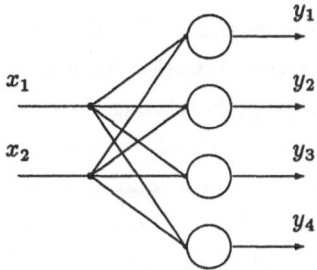

Figure B.6: Self-organizing behavior.

the updating law for the weights to be

$$w_{ij_{new}} = w_{ij_{old}} + .5(x_j - w_{ij_{old}})y_i.$$

Notice that the only two weights that will be adjusted on each trial are those associated with the neuron with the smallest distance. For this reason we call this a competitive learning algorithm. Now suppose we start off with the weights set to some arbitrary random values and we present the following sequence of input vectors over and over:

$$\begin{pmatrix} 0 \\ 0 \end{pmatrix} \rightarrow \begin{pmatrix} 0 \\ 1 \end{pmatrix} \rightarrow \begin{pmatrix} 1 \\ 0 \end{pmatrix} \rightarrow \begin{pmatrix} 1 \\ 1 \end{pmatrix} \rightarrow \begin{pmatrix} 0 \\ 0 \end{pmatrix} \cdots,$$

adjusting the weights according to the formula given above. What we will find is that eventually the weight vectors converge such that there is one weight vector identically matched to each input vector. Thus, the net will light up a single output for each input vector whenever that vector is presented. This is a simple example of self-organization, however, it illustrates the basic idea used to develop more complicated examples and applications.

B.3.3 Optimization

ANNs can also be used to solve optimization problems. Actually, this approach is an extension of the associative memory properties. It has been shown that the trajectory of a Hopfield net moves to minimize an energy function of the system. Typically, the stable equilibrium points to which the system's trajectories converge are local minimums of the system's energy function. For Hopfield-type nets, it is possible to define an energy function for the network in terms of the interconnection weights of the network. This fact can be used to solve some optimization problems, by properly relating the quantity you wish to optimize to the weights of the net, which are related to the energy function of the neural net.

An example of this was Hopfield's solution of the traveling salesman problem using a neural network [105]. This is a classic "hard" optimization problem. A traveling salesman has the task of visiting each of N cities, without visiting any city twice. The distance between each city is known and his problem is to figure out what route he should follow to make his trip as short as possible. Hopfield showed how to set up the weights of a neural network in terms of the distances between the cities so that the problem could be solved, at least for a relatively small number of cities.

Another example of using an ANN to solve optimization problems is the result of Kennedy and Chua [73]. They have shown how to set up the weights of a neural net to solve the general nonlinear programming problem. This is a very important result, because many engineering problems can be cast as nonlinear programs. Consider the problem

$$\min_{v \in R^n} \phi(v),$$

subject to

$$f_i(v) \geq 0,$$

for $i = 1, \ldots, q$.

Kennedy and Chua were able to show that the canonical nonlinear programming circuit shown in Figure B.7 solves this nonlinear programming problem (under standard assumptions of constraint qualification, continuity, existence of first and second partials, etc.). Although it may not be obvious, the circuit is a Hopfield neural network, described by a set of coupled equations of the form

$$C_i \frac{dv_i}{dt} = -\frac{\partial \phi}{\partial v_i} - \sum_{j=1}^{p} g(f(v)) \frac{\partial f_j}{\partial v_i},$$

for $i = 1, \ldots, q$, where $g(\cdot)$ is a passive monotone nondecreasing function.

128

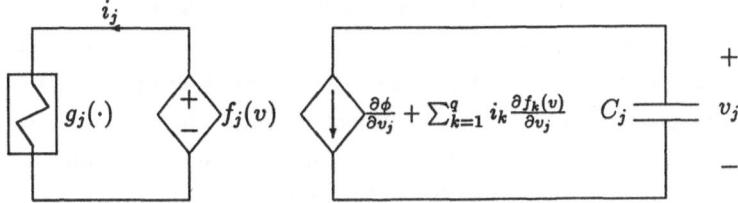

Figure B.7: Neural network for optimization.

B.3.4 Nonlinear Mappings

In 1900, mathematician David Hilbert posed what became known as his 13th problem. The problem dealt with whether or not a real-valued function could always be broken down into the sum of a finite number of other functions of a fewer number of variables. Hilbert conjectured that the answer was negative. In 1957 the Russian mathematician Kolmogorov disproved Hilbert's conjecture with his so-called superposition theorem. This theorem has direct application to the field of neural nets, and implies that any function from R^n to R^m can be implemented or realized with a feedforward neural network (subject to certain restrictions, of course). However, this important theoretical result is an existence condition and it is not yet known how to select the weights *a priori* to implement a given mapping with a feedforward network. Nevertheless, many researchers are studying ways to learn nonlinear mappings with neural nets, because the applications of such a result would be far-reaching. As one example, consider the control field. Because most real systems are nonlinear, it often makes sense to control them using a nonlinear controller. However, the analysis and design of nonlinear control systems is very hard. Hence, a neural net that could learn the correct nonlinear control action through training would be very useful, particularly for applications in robotics and other motion control systems. For further discussion about the use of neural networks for control system applications, refer back to the first section of Chapter 7. Also see the recent book, *Neural Networks for Control* [65].

As we have seen, a key ingredient governing the net's behavior is its learning law. To exploit the ability of an ANN to implement a nonlinear mapping (which is really a generalization of associative memory, pattern recognition, and self-organization properties), it is necessary that we have a good learning law.

Most learning algorithms currently being studied, such as backpropagation, are gradient descents in the weight space. This is one approach to what has been called the "credit assignment" problem. That is, how to reinforce weights that contribute to desirable behavior of the system. Another approach to the credit assignment problem is called reinforcement learning. In this approach we use a single reinforcement signal from the environment (system) to decide how to adjust the weights. This is decidedly different from backpropagation, which essentially attempts to form an error signal at the output of each neuron in the network. In reinforcement learning we have only one error signal, which is used by all the neurons.

Reinforcement learning can be used to help teach a neural network a desired nonlinear control action. To describe reinforcement learning, consider Figure B.8. Here we suppose the neural net generates control actions, based on measurements

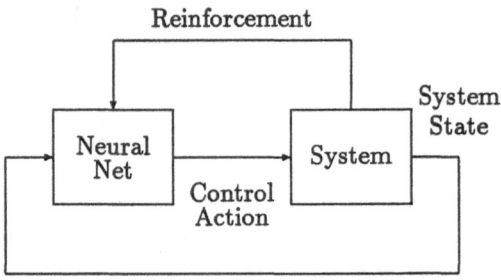

Figure B.8: Reinforcement learning.

of the current system state. It then receives a reinforcement signal from the environment that evaluates the effect of the action. The weights are adjusted based upon the reinforcement. In the learning algorithm, the only information given to individual neurons, regardless of their location in the net, is the reinforcement signal. There is no error that is backpropagated through the net. In a typical algorithm, only those weights that contributed to the action of the net will be adjusted. This can be achieved by associating an eligibility function with each weight. This function indicates how recently the weight has been changed or how recently the neurons associated with a weight have fired. Only weights that are "eligible" are updated during learning. Learning initially proceeds randomly, but as the net learns which actions lead to a desirable reinforcement in response to specific system states, the weights begin to adjust in such a way that a bias is built up toward these actions. This type of system is in fact a stochastic automata. Past work in reinforcement learning drew from Kampati Narendra's work in stochastic automata theory [26] and Klopf's studies in psychology [106]. More recently, a research group based around Andrew Barto, Richard Sutton and Charles Anderson has been very active. In what has become a very well-known paper [78], they used reinforcement learning to solve the classic "broomstick-balancing" problem. (Incidentally, Widrow also solved this problem using an ADALINE network in the early 1960's.) Their solution used a network made up of what they called "neuron-like" elements, which were basically perceptrons with stochastic outputs. The network applied control forces to a cart-pole system in an attempt to balance the pole. The system was able to learn which actions would keep the pole balanced by adjusting the weights whenever the pole fell over (negative reinforcement). These ideas have been applied to robotic manipulator control [80], to nuclear reactor heat-up control [107], and to the control of aerospace vehicles [83].

B.4 NEURAL NETS AND COMPUTERS

One question that arises in discussions about neural networks concerns their relationship to conventional computers. In the early days of computers, people referred to them as "electronic brains." Research has now shown us that this is an outdated paradigm. It is perhaps more accurate to refer to neural net-based systems as being brainlike, rather than the conventional digital computer. There are several reasons for this. One, the biological neuron is in fact an analog device and many ANN properties result from analog models. The computer is by definition digital. Second, computers work on a nanosecond time scale. The neurons of the brain, on the other hand, have time constants on the order of 100 milliseconds. A third note is that a conventional von Neumann-type computer uses highly complex software, yet the ANN has a low degree of functional complexity. For example, the code a computer must run to recognize an image is enormous, and very susceptible to noise and faulty data. Such software may take seconds of CPU time to perform a simple pattern recognition problem. However, a biological neural net can solve the same problem within one or two time constants, without a complex program. A fourth comparison is that a computer is a sequential device, which does one thing at a time. In contrast, biological neural nets are characterized by massive parallelism. In fact, neural net approaches are often called connectionist approaches. This is because the ANNs are motivated by the biological neural nets, which appear to derive their computational capabilities from their high degree of connectivity, rather than from functional complexity. The idea is to hook up a large number of identical processing elements, each performing essentially the same computations. One might note here that the examples presented in this monograph have not included a high degree of connectivity. This is true. In fact, many of the research examples in the literature deal with a small number of neurons, and when the number of neurons increase, the results don't always work as well computationally. However, this is a research field and all the important work is not done yet.

Another point is that the comparisons we just made do not indict the digital computer. We must allow that there are some tasks that computers are better suited to do and some tasks that neural nets are better suited to do. Because neural nets exhibit some of the same properties that biological systems exhibit, they often suffer the same limitations. Just as a computer does a better job than a person at balancing a checkbook (fewer errors, faster), it will probably also do better at this job than a neural net. Similarly, the neural net will do better at complex control and pattern recognition tasks. In fact, people are already realizing these facts and are developing "hybrid" systems, which combine digital computers with neural nets.

B.5   DERIVATION OF BACKPROPAGATION

In this section we derive the backpropagation learning rule for the neural network shown in Figure B.2. Although the results of this derivation are available in the literature, for completeness we present the details here, for the special case of a feedforward net with two hidden layers.

As noted above, the basic idea of the algorithm is to change the weights in the direction of decreasing energy, where energy means some measure of the errors in the neural network's outputs. Suppose we want the system's outputs to be y_k^d, where the subscript denotes the kth element of the desired output vector $\underline{y}_d$. Define the error $\delta y_k = (y_k^d - y_k)$. The energy we wish to minimize is the mean-square error, given by

$$E = \frac{1}{2} \sum_{k=1}^{N} (\delta y_k)^2 = \frac{1}{2} \sum_{k=1}^{N} (y_k^d - y_k)^2.$$

132

We then choose weight changes according to

$$\Delta w_{ki} \propto -\frac{\partial E}{\partial w_{ki}},$$

$$\Delta \alpha_{ij} \propto -\frac{\partial E}{\partial \alpha_{ij}},$$

$$\Delta \beta_{jl} \propto -\frac{\partial E}{\partial \beta_{jl}}.$$

This will be done by straightforward application of the chain rule.

• *Output layer weights*

We begin with the weights w_{ki}. Writing

$$
\begin{aligned}
\frac{\partial E}{\partial w_{ki}} &= \frac{\partial}{\partial w_{ki}} [\frac{1}{2} \sum_{k=1}^{N} (y_k^d - y_k)^2] \\
&= -\sum_{k=1}^{N} \delta y_k \frac{\partial y_k}{\partial w_{ki}} \\
&= -\sum_{k=1}^{N} \delta y_k \frac{\partial}{\partial w_{ki}} [\sum_{i=1}^{M} w_{ki} v_i] \\
&= -\delta y_k v_i,
\end{aligned}
$$

and so we have

$$\Delta w_{ki} = \eta_1 \delta y_k v_i,$$

where η_1 is a proportionality constant.

• *Second hidden-layer weights*

Next consider the weights α_{ij}. Computing the negative gradient of the error, we again have

$$\frac{\partial E}{\partial \alpha_{ij}} = -\sum_{k=1}^{N} \delta y_k \frac{\partial y_k}{\partial \alpha_{ij}}.$$

To evaluate this, compute

$$
\begin{aligned}
\frac{\partial y_k}{\partial \alpha_{ij}} &= \frac{\partial}{\partial \alpha_{ij}} [\sum_{i=1}^{M} w_{ki} f(\sum_{j=1}^{M} \alpha_{ij} z_j)] \\
&= \sum_{i=1}^{M} w_{ki} \frac{\partial}{\partial \alpha_{ij}} [f(\sum_{j=1}^{M} \alpha_{ij} z_j)]
\end{aligned}
$$

133

$$= \sum_{i=1}^{M} w_{ki} f'(\sum_{j=1}^{M} \alpha_{ij} z_j) \frac{\partial}{\partial \alpha_{ij}} [\sum_{j=1}^{M} \alpha_{ij} z_j]$$

$$= w_{ki} f'(\sum_{j=1}^{M} \alpha_{ij} z_j) z_j.$$

Now, note that for the sigmoid function

$$f(x) = \frac{1}{1 + e^{-ax}},$$

we have

$$\frac{d}{dx} f(x) = a f(x)[1 - f(x)].$$

Also, by definition of the network

$$v_i = f(\sum_{j=1}^{M} \alpha_{ij} z_j),$$

which means that

$$f'(\sum_{j=1}^{M} \alpha_{ij} z_j) = a v_i (1 - v_i).$$

This gives

$$\frac{\partial y_t}{\partial \alpha_{ij}} = w_{ki}[a v_i (1 - v_i)] z_j,$$

and thus,

$$\frac{\partial E}{\partial \alpha_{ij}} = -\sum_{k=1}^{N} \delta y_k w_{ki}[a v_i (1 - v_i)] z_j$$

$$= -a v_i (1 - v_i)[\sum_{k=1}^{N} w_{ki} \delta y_k] z_j.$$

Finally, if we define

$$\delta v_i = a v_i (1 - v_i) \sum_{k=1}^{N} w_{ki} \delta y_k,$$

then we obtain

$$\Delta \alpha_{ij} = \eta_2 \delta v_i z_j.$$

134

• *First hidden layer weights*

Now consider the weights β_{jl}. Computing the gradient, we have

$$\frac{\partial E}{\partial \beta_{jl}} = -\sum_{k=1}^{N} \delta y_k \frac{\partial y_t}{\partial \beta_{jl}}.$$

To evaluate this, compute

$$
\begin{aligned}
\frac{\partial y_t}{\partial \beta_{jl}} &= \frac{\partial}{\partial \beta_{jl}} [\sum_{i=1}^{M} w_{ki} f(\sum_{j=1}^{M} \alpha_{ij} f(\sum_{l=1}^{N} \beta_{jl} u_l))] \\
&= \sum_{i=1}^{M} w_{ki} f\prime(\sum_{j=1}^{M} \alpha_{ij} z_j) \frac{\partial}{\partial \beta_{jl}} [\sum_{j=1}^{M} \alpha_{ij} f(\sum_{l=1}^{N} \beta_{jl} u_l)] \\
&= \sum_{i=1}^{M} w_{ki} [a v_i (1 - v_i)] \sum_{j=1}^{M} \alpha_{ij} f\prime(\sum_{l=1}^{N} \beta_{jl} u_l) \frac{\partial}{\partial \beta_{jl}} [\sum_{l=1}^{N} \beta_{jl} u_l] \\
&= \sum_{i=1}^{M} w_{ki} [a v_i (1 - v_i)] \alpha_{ij} [a z_j (1 - z_j)] u_l.
\end{aligned}
$$

So the gradient becomes

$$
\begin{aligned}
\frac{\partial E}{\partial \beta_{jl}} &= -\sum_{k=1}^{N} \delta y_k \sum_{i=1}^{M} w_{ki} [a v_i (1 - v_i)] \alpha_{ij} [a z_j (1 - z_j)] u_l \\
&= -a z_j (1 - z_j) \{\sum_{i=1}^{M} \alpha_{ij} [a v_i (1 - v_i) \sum_{k=1}^{N} w_{ki} \delta y_k]\} u_l \\
&= -a z_j (1 - z_j) [\sum_{i=1}^{M} \alpha_{ij} \delta v_i] u_l.
\end{aligned}
$$

Therefore, if we define

$$\delta z_j = a z_j (1 - z_j) \sum_{i=1}^{M} \alpha_{ij} \delta v_i,$$

then we obtain

$$\Delta \beta_{jl} = \eta_3 \delta z_j u_l.$$

B.6 NEURAL NETWORK REFERENCES

Our overview of the field of neural nets has been cursory, to say the least. There is much more to know, from stochastic neural nets called Boltzman machines, which learn using simulated annealing, to neural nets with fuzzy logic, to bidirectional associative memories (BAM's), to neocognitrons, to counter-propagation nets (CPN), to neural nets that exhibit chaotic behavior, etc.

As we have mentioned, the literature related to neural networks is exhaustive (or perhaps we should say exhausting!). For those with a lot of stamina, however, we can suggest a few things to read. Two introductory articles are Hecht-Nielsen's "Neurocomputing: Picking the human brain," [108] and Lippman's survey, "An Introduction to Computing with Neural Nets," [103]. Hecht-Nielsen's paper is a good overview of the field and also describes some of the hardware available to do neurocomputing. Lippman's paper is already considered a classic and is an indispensable reference, particularly for electrical engineers.

To delve into the research literature there are a number of sources. There are several refereed journals that publish research results in the field. The International Neural Networks Society (INNS) has published the journal *Neural Networks* since 1988. There is also a journal called *Neural Computation* that began in 1989. Also, the IEEE began publishing the *IEEE Transactions on Neural Networks* in 1990. Another good source of research papers is the proceedings of the International Joint Conference on Neural Networks (IJCNN), sponsored by the INNS and the IEEE. This conference started in 1987 as the IEEE ICNN, and a few years ago became the IJCNN. A particularly good source of application papers can be found in the April issues (1988–1990, 1992) of the *IEEE Control Systems Magazine*. See also special issues of the *IEEE Transactions on Circuit and Systems* (May 1989) and the *IEEE Transactions on ASSP* (July 1988). You can also find articles on neural networks in almost any journal or magazine you pick up.

In addition to articles and papers, there are some books you should have on your reference list. We have mentioned Hecht-Nielsen's book, *Neurocomputing* [99]. It is written as a senior-level/graduate-level text and was one of the first textbooks available. Bark Kosko has recently published a textbook titled *Neural Networks and Fuzzy Systems* [109]. This book takes a dynamical system approach to neural networks. We previously referred to the PDP book [100]. Sections of this are required reading for people seriously interested in the field. The standard reference related to self-organization is Kohonen's book, *Self-Organization and Associative Memory* [70]. Four books of historical interest are Minsky and Papert's *Perceptrons* (1988 expanded edition) [98], Nils Nilsson's 1965 book, *The Mathematical Foundations of Learning Machines* (revised 1990) [110], Rosenblatt's 1961 book, *Principles of Neurodynamics* [111], and Hebb's 1949 book, *The Organization of Behavior* [95]. Two final sources to recommend are *Neurocomputing, Foundations of Research* [112] and *Neurocomputing 2, Directions for Research* [113], both edited by Anderson and Rosenfeld. The first of these is a collection of classic papers, all the way back to McCollough and Pitts. The second is a collection of more recent results. Another collection is an IEEE Computer Society Press publication, *Artificial Neural Networks: Theoretical Concepts*, edited by V. Vemuri [114]. In the time it takes this monograph to come to press it is certain that some new books and papers will appear. However, this list will give you a good start.

REFERENCES

[1] T. Kailath, *Linear Systems*. Englewood Cliffs, New Jersey: Prentice-Hall, 1980.

[2] D. Ohm, J. Howze, and S. Bhattacharyya, "Structural synthesis of multivariable controllers," *Automatica*, vol. 21, pp. 35–55, January 1985.

[3] M. Dahleh and J. Pearson, "Minimization of a regulated response to a fixed input," *IEEE Transactions on Automatic Control*, vol. 33, pp. 924–930, October 1988.

[4] K. Moore and S. Bhattacharyya, "A technique for choosing zero locations for minimal overshoot," *IEEE Transactions on Automatic Control*, vol. 35, pp. 577–580, May 1990.

[5] T. Mita and Y. Chida, "2-delay digital robust control – avoiding the problem of unstable zeros," in *Proceedings of 27th Conference on Decision and Control*, (Austin, Texas), pp. 1883–1888, December 1988.

[6] K. Moore, S. Bhattacharyya, and M. Dahleh, "Arbitrary pole and zero assignment with N-delay input control using stable controllers," in *Proceedings of 28th IEEE Conference on Decision and Control*, (Tampa, Florida), December 1989.

[7] P. Bondi, G. Casalino, and L. Gambardella, "On the iterative learning control theory for robotic manipulators," *IEEE Journal of Robotics and Automation*, vol. 4, pp. 14–22, February 1988.

[8] M. Uchiyama, "Formation of high speed motion pattern of mechanical arm by trial," *Transactions of the Society of Instrumentation and Control Engineers*, vol. 19, pp. 706–712, May 1978.

[9] S. Arimoto, S. Kawamura, and F. Miyazaki, "Bettering operation of robots by learning," *Journal of Robotic Systems*, vol. 1, pp. 123–140, March 1984.

[10] S.Arimoto, S. Kawamura, and F.Miyazaki, "Can mechanical robots learn by themselves?," in *Proceedings of 2nd International Symposium of Robotics Research*, (Kyoto, Japan), pp. 127–134, August 1984.

[11] S. Arimoto, S. Kawamura, and F. Miyazaki, "Iterative learning control for robot systems," in *Proceedings of IECON*, (Tokyo, Japan), October 1984.

[12] S. Arimoto, S. Kawamura, and F. Miyazaki, "Bettering operation of dynamic systems by learning: a new control theory for servomechanism or mechatronic systems," in *Proceedings of 23rd Conference on Decision and Control*, (Las Vegas, Nevada), pp. 1064–1069, December 1984.

[13] S. Kawamura, F. Miyazaki, and S. Arimoto, "Hybrid position/force control of robot manipulators based on learning method," in *Proceedings of '85 International Conference on Advanced Robotics*, (Tokyo, Japan), pp. 235–242, September 1985.

[14] S. Arimoto, "Mathematical theory of learning with applications to robot control," in *Proceedings of 4th Yale Workshop on Applications of Adaptive Systems*, (New Haven, Connecticut), pp. 379–388, May 1985.

[15] S. Arimoto, "System theoretic study of learning control," *Transactions of Society of Instrumentation and Control Engineers*, vol. 21, no. 5, pp. 445–450, 1985.

[16] S. Arimoto, S. Kawamura, F. Miyazaki, and S. Tamaki, "Learning control theory for dynamic systems," in *Proceedings of 24th Conference on Decision and Control*, (Ft. Lauderdale, Florida), pp. 1375–1380, December 1985.

[17] S. Kawamura, F. Miyazaki, and S. Arimoto, "Applications of learning method for dynamic control of robot manipulators," in *Proceedings of 24th Conference on Decision and Control*, (Ft. Lauderdale, Florida), pp. 1381–1386, December 1985.

[18] S. Kawamura, F. Miyazaki, and S. Arimoto, "Convergence, stability and robustness of learning control schemes for robot manipulators," in *Proceedings of the International Symposium on Robotic Manipulators: Modelling, Control, and Education*, (Albuquerque, New Mexico), November 1986.

[19] S. Kawamura, F. Miyazaki, M. Matsumori, and S. Arimoto, "Learning control scheme for a class of robot systems with elasticity," in *Proceedings of*

25th *Conference on Decision and Control*, (Athens, Greece), pp. 74–79, December 1986.

[20] S. Kawamura, F. Miyazaki, and S. Arimoto, "Intelligent control of robot motion based on learning method," in *Proceedings of IEEE International Symposium on Intelligent Control*, (Philadelphia, Pennsylvania), pp. 365–370, January 1987.

[21] S. Kawamura, F. Miyazaki, and S. Arimoto, "Realization of robot motion based on a learning method," *IEEE Transactions on Systems, Man and Cybernetics*, vol. 18, pp. 126–134, January/February 1988.

[22] Y. Tsypkin, *Adaptation and Learning in Automatic Systems*. New York, New York: Academic Press, 1971.

[23] Y. Tsypkin, *Foundations of the Theory of Learning Systems*. New York, New York: Academic Press, 1973.

[24] K. Fu, "Learning control systems — review and outlook," *IEEE Transactions on Automatic Control*, vol. 15, pp. 210–215, April 1970.

[25] AACC Theory Committee, *Learning Systems*. Columbus, Ohio: Automatic Control Council, 1973.

[26] K. Narendra and M. Thathachar, "Learning automata — a survey," *IEEE Transactions on Systems, Man, and Cybernetics*, vol. 4, pp. 323–334, July 1974.

[27] G. Sardidis, *Self-Organizing Control of Stochastic Systems*. New York, New York: Marcel Dekker, 1977.

[28] J. Plant, *Some Iterative Solutions in Optimal Control*. Cambridge, Massachusetts: M.I.T. Press, 1968.

[29] R. J. LaCarna and J. R. Johnson, "A learning controller for the megawatt load-frequency control problem," *IEEE Transactions on Systems, Man, and Cybernetics*, vol. 10, pp. 43–49, January 1980.

[30] T. Mita and E. Kato, "Iterative control and its application to motion control of robot arm – a direct approach to servo-problems," in *Proceedings of 24th Conference on Decision and Control*, (Ft. Lauderdale, Florida), pp. 1393–1398, December 1985.

[31] T. Mita, "Repetitive control of mechanical systems," in *Proceedings of 1984 ATACS*, (Izu, Japan), 1984.

[32] S. Hara, T. Omata, and M. Nakano, "Synthesis of repetitive control systems and its application," in *Proceedings of 24th Conference on Decision and Control*, (Ft. Lauderdale, Florida), pp. 1387–1392, December 1985.

[33] S. Hara, Y. Yamamoto, T. Omata, and M. Nakano, "Repetitive control system: a new type servo system for periodic exogenous signals," *IEEE Transactions on Automatic Control*, vol. 33, pp. 659–668, July 1988.

[34] M. Togai and O. Yamano, "Analysis and design of an optimal learning control scheme for industrial robots: a discrete system approach," in *Proceedings of 24th Conference on Decision and Control*, (Ft. Lauderdale, Florida), pp. 1399–1404, December 1985.

[35] K. Furuta and M. Yamakita, "The design of a learning control system for multivariable systems," in *Proceedings of IEEE International Symposium on Intelligent Control*, (Philadelphia, Pennsylvania), pp. 371–376, January 1987.

[36] C. Atkeson and J. McIntyre, "Robot trajectory learning through practice," in *Proceedings of IEEE Conference on Robotics and Automation*, (San Francisco, California), April 1986.

[37] L. Hideg and R. Judd, "Frequency domain analysis of learning systems," in *Proceedings of the 27th Conference on Decision and Control*, (Austin, Texas), pp. 586–591, December 1988.

[38] S. Oh, Z. Bien, and I. Suh, "An iterative learning control method with application for the robot manipulator," *IEEE Journal of Robotics and Automation*, vol. 4, pp. 508–514, October 1988.

[39] J. Craig, "Adaptive control of manipulators through repeated trials," in *Proceedings of 1984 American Control Conference*, (San Diego, California), pp. 1566–1572, June 1984.

[40] J. Craig, *Adaptive Control of Robot Manipulators*. Reading, Massachusetts: Addison-Wesley, 1988.

[41] J. Craig, P. Hsu, and S. Sastry, "Adaptive control of mechanical manipulators," in *Proceedings of IEEE Conference on Robotics and Automation*, (San Francisco, California), pp. 190–195, April 1986.

[42] Y. Gu and N. Loh, "Learning control in robotic systems," in *Proceedings of IEEE International Symposium on Intelligent Control*, (Philadelphia, Pennsylvania), pp. 360–364, January 1987.

[43] E. Harokopos, "Optimal learning control of mechanical manipulators in repetitive motions," in *Proceedings of IEEE International Symposium on Intelligent Control*, (Philadelphia, Pennsylvania), pp. 396–401, January 1987.

[44] M. Yamakita and K. Furuta, "Iterative generation of virtual reference for a manipulator," *Robotica*, vol. 9, pp. 71–80, 1991.

[45] Z. Bien, D. Hwang, and S. Oh, "A nonlinear iterative learning method for robot path control," *Robotica*, vol. 9, pp. 387–392, 1991.

[46] W. Messner, R. Horowitz, W. Kao, and M. Boals, "A new adaptive learning rule," *IEEE Transactions on Automatic Control*, vol. 36, pp. 188–197, February 1991.

[47] R. Horowitz, W. Messner, and J. B. Moore, "Exponential convergence of a learning controller for robot manipulators," *IEEE Transactions on Automatic Control*, vol. 36, pp. 890–894, July 1991.

[48] S. Wang, "Computed reference error adjustment technique (CREATE) for the control of robot manipulators," in *Proceedings of 22nd Annual Allerton Conference*, pp. 874–876, October 1984.

[49] S.Wang, "Inversion of nonlinear dynamical systems," Technical Report, Dept. of Electrical and Computer Engineering, University of California, Davis, California, 1988.

[50] S. Wang and I. Horowitz, "CREATE – a new adaptive technique," in *Proceedings of 19th Annual Conference on Information Sciences and Systems*, March 1985.

[51] J. Hauser, "Learning control for a class of nonlinear systems," in *Proceedings of 26th Conference on Decision and Control*, (Los Angeles, California), pp. 859–860, December 1987.

[52] G. Heinzinger, D. Fenwick, B. Paden, and F. Miyazaki, "Robust learning control," in *Proceedings of the 28th Conference on Decision and Control*, (Tampa, Florida), pp. 436–440, December 1989.

[53] T. Sugie and T. Ono, "An iterative learning control law for dynamical systems," *Automatica*, vol. 27, pp. 729–732, 1991.

[54] J. Hodgson, "Structures for intelligent systems," in *Proceedings of IEEE International Symposium on Intelligent Control*, (Philadelphia, Pennsylvania), pp. 348–353, January 1987.

[55] Y. Suganuma and M. Ito, "Learning control and knowledge representation," in *Proceedings of IEEE International Symposium on Intelligent Control*, (Philadelphia, Pennsylvania), pp. 354–359, January 1987.

[56] M. Kawato, K. Furukawa, and R. Suzuki, "A hierarchical neural-network model for control and learning of voluntary movement," *Biological Cybernetics*, vol. 57, pp. 169–185, February 1987.

[57] B. Francis, *A Course in H_∞ Control Theory, Lecture Notes in Control and Information Sciences*. Berlin: Springer-Verlag, 1987.

[58] K. Glover, "All optimal Hankel-norm approximations of linear multivariable systems and their L^∞-error norms," *International Journal of Control*, vol. 39, no. 6, pp. 1115–1193, 1984.

[59] M. Dahleh and J. Pearson, "L^1-optimal compensators for continuous-time systems," *IEEE Transactions on Automatic Control*, vol. 32, pp. 889–895, October 1987.

[60] G. Goodwin and K. Sin, *Adaptive Filtering, Prediction and Control*. Englewood Cliffs, New Jersey: Prentice-Hall, 1984.

[61] C. Lawson and R. Hanson, *Solving Least Squares Problems*. Englewood Cliffs, New Jersey: Prentice-Hall, 1974.

[62] B. Kuo, *Automatic Control Systems*. Englewood Cliffs, New Jersey: Prentice-Hall, 4th ed., 1984.

[63] K. Moore, *Design Techniques for Transient Response Control*. Ph.D. thesis, Texas A&M University, College Station, Texas, 1989.

[64] K. Narendra and K. Parthasarathy, "Identification and control of dynamical systems using neural networks," *IEEE Transactions on Neural Networks*, vol. 1, pp. 4–27, March 1990.

[65] W. T. Miller, R. S. Sutton, and P. J. Werbos, *Neural Networks for Control.* Cambridge, Massachusetts: The MIT Press, 1990.

[66] K. Moore, "A control-theoretic perspective on learning in neural networks," Final Report: Part I, 1991 AWU-DOE Faculty Fellowship Program, INEL/EG&G Idaho, Science and Technology Department, Materials Technology Group, Metals Joining Section, Idaho Falls, Idaho, August 1991.

[67] A. Guez, J. Eibert, and M. Cam, "Neural network architecture for control," *IEEE Control Systems Magazine*, vol. 8, pp. 22–24, April 1988.

[68] A.Guez, V. Protopopsecu, and J. Barhen, "On the stability, storage capacity, and design of nonlinear continuous neural networks," *IEEE Transactions on Systems, Man, and Cybernetics*, vol. 18, pp. 80–87, January/February 1988.

[69] S. Kumar and A. Guez, "ART-based adaptive pole placement for neurocontrollers," *Neural Networks*, vol. 4, pp. 319–335, 1991.

[70] T. Kohonen, *Self-Organization and Associative Memory, Lecture Notes in Control and Information Theory.* Berlin: Springer-Verlag, 1986.

[71] T. Kohonen, "The self-organizing map," *Proceedings of the IEEE*, vol. 78, pp. 1464–1480, September 1990.

[72] D. Tank and J. Hopfield, "Simple neural optimization networks: an A/D converter, signal decision circuit, and a linear programming circuit," *IEEE Transactions on Circuits and Systems*, vol. 33, pp. 533–541, May 1986.

[73] M. Kennedy and L. Chua, "Neural networks for nonlinear programming," *IEEE Transactions on Circuits and Systems*, vol. 35, pp. 554–562, May 1988.

[74] K. Moore and S. Naidu, "Linear quadratic regulation using neural networks," in *Proceedings of the 1991 International Joint Conference on Neural Networks*, (Seattle, Washington), August 1991.

[75] R. Hecht-Nielsen, "Kolmogorov's mapping neural network existence theorem," in *Proceedings of IEEE 1987 International Conference on Neural Networks*, (San Diego, California), pp. 1–4, June 1987.

[76] R. D. Figueiredo, "Implications and applications of Kolmogorov's superposition theorem," *IEEE Transactions on Automatic Control*, vol. 25, pp. 1227–1231, December 1980.

[77] D. Psaltis, A. Sideris, and A. Yamamura, "A multilayered neural network controller," *IEEE Control Systems Magazine*, vol. 8, pp. 17–21, April 1988.

[78] A. Barto, R. Sutton, and C. Anderson, "Neuron-like adaptive elements that can solve difficult learning control problems," *IEEE Transactions on Systems, Man, and Cybernetics*, vol. 13, pp. 834–846, September 1983.

[79] V. Gullapalli, "A stochastic reinforcement learning algorithm for learning real-valued functions," *Neural Networks*, vol. 3, pp. 671–692, 1990.

[80] J. A. Franklin, "Refinement of robot motor skills through reinforcement learning," in *Proceedings of 27th Conference on Decision and Control*, (Austin, Texas), pp. 1096–1101, December 1988.

[81] C. W. Anderson, "Strategy learning with multilayer connectionist representations," GTE Labs Technical Report TR87-509.3, GTE Laboratories, Waltham, Massachusetts, May 1988.

[82] R. Sutton, "Learning to predict by the methods of temporal differences," *Machine Learning*, vol. 3, pp. 9–44, 1988.

[83] K. Moore, "Neural networks for guidance, navigation, and control of exoatmospheric interceptors," Final Report, 1990 Summer Faculty Fellowship Program, AFATL/SAI, Air Force Armament Laboratory, Eglin AFB, Florida, September 1990.

[84] O. Sebakhy, M. El-Singaby, and I. El-Arabawy, "Shaping the output response in a class of linear multivariable systems," *IEEE Transactions on Automatic Control*, vol. 33, pp. 457–458, May 1988.

[85] J. Doyle, B. A. Francis, and A. R. Tannenbaum, *Feedback Control Systems*. New York, New York: Macmillan Publishing Company, 1992.

146

[86] T. Hagiwara and M. Araki, "Design of stable state feedback controller based on multirate sampling of the plant output," *IEEE Transactions on Automatic Control*, vol. 33, pp. 812–819, September 1988.

[87] B. Francis and T. Georgiou, "Stability theory for linear time-invariant plants with periodic digital controllers," *IEEE Transactions on Automatic Control*, vol. 33, pp. 820–832, September 1988.

[88] K. Ogata, *Discrete-Time Control Systems*. Englewood Cliffs, New Jersey: Prentice-Hall, 1987.

[89] E. Jury, *Sampled-Data Control Systems*. New York, New York: John Wiley and Sons, 1958.

[90] C. Chen, *Linear System Theory and Design*. New York, New York: Holt, Rinehart and Wilson, 1984.

[91] J. Hopfield, "Neural networks and physical systems with emergent collective computational abilities," *Proceedings of the National Academy of Science*, vol. 79, pp. 2554–2558, April 1982.

[92] J. Hopfield, "Neurons with graded response have collective computational properties like those of two-state neurons," *Proceedings of the National Academy of Science*, vol. 81, pp. 3088–3092, May 1984.

[93] M. Cohen and S. Grossberg, "Absolute stability of global pattern formation and parallel memory storage by competitive neural networks," *IEEE Transactions on Systems, Man, and Cybernetic*, vol. 13, pp. 815–826, September/October 1983.

[94] W. S. McCulloch and W. Pitts, "A logical calculus of ideas immanent in nervous activity," *Bulletin of Mathematical Biophysics*, vol. 5, pp. 115–133, 1943.

[95] D. Hebb, *The Organization of Behavior*. New York, New York: John Wiley and Sons, 1949.

[96] F. Rosenblatt, "The perceptron: a probabilistic model for information storage and organization in the brain," *Psychological Review*, vol. 65, pp. 386–408, 1958.

[97] B. Widrow, R. Winter, and R. Baxter, "Layered neural nets for pattern recognition," *IEEE Transactions on Acoustics, Speech, and Signal Processing*, vol. 36, pp. 1109–1118, July 1988.

[98] M. L. Minsky and S. A. Papert, *Perceptrons*. Cambridge, Massachusetts: The MIT Press, 1988.

[99] R. Hecht-Nielsen, *Neurocomputing*. Reading, Massachusetts: Addison-Wesley, 1990.

[100] D. Rumelhart and J. McClelland, *Parallel Distributed Processing: Explorations in the Microstructure of Cognition, vols. I and II*. Cambridge, Massachusetts: MIT Press, 1986.

[101] P. J. Werbos, *Beyond Regression: New Tools for Prediction and Analysis in the Behavioral Sciences*. Ph.D. thesis, Harvard University, Cambridge, Massachusetts, 1974.

[102] D. B. Parker, "Learning logic," Technical Report TR-47, Center for Computational Research in Economics and Management Science, MIT, Cambridge, Massachusetts, April 1985.

[103] R. Lippman, "An introduction to computing with neural networks," *IEEE ASSP Magazine*, vol. 4, pp. 4–22, April 1987.

[104] A. N. Michel and J. A. Farrell, "Associative memories via artificial neural networks," *IEEE Control Systems Magazine*, vol. 10, pp. 6–17, April 1990.

[105] J. Hopfield and D. Tank, "Neural computations of decisions in optimization problems," *Biological Cybernetics*, vol. 52, pp. 141–152, 1985.

[106] A. Klopf, *The Hedonistic Neuron: A Theory of Memory, Learning, and Intelligence*. Washington, D.C.: Hemisphere, 1982.

[107] W. Jouse and J. Williams, "PWR Heat-Up Control by Means of a Self-Teaching Neural Network," in *Proceedings of International Conference on Control and Instrumentation in Nuclear Installations*, (Glasgow), May 1990.

[108] R. Hecht-Nielsen, "Neurocomputing: picking the human brain," *IEEE Spectrum*, vol. 25, pp. 36–41, March 1988.

[109] B. Kosko, *Neural Networks and Fuzzy Systems*. Englewood Cliffs, New Jersey: Prentice Hall, 1992.

[110] N. Nilsson, *The Mathematical Foundations of Learning Machines*. San Mateo, California: Morgan Kaufmann Publishers, 1990.

[111] F. Rosenblatt, *Principles of Neurodynamics*. Washington, D.C.: Spartan Books, 1961.

[112] J. A. Anderson and E. Rosenfeld, *Neurocomputing: Foundations of Research*. Cambridge, Massachusetts: The MIT Press, 1988.

[113] J. A. Anderson and E. Rosenfeld, *Neurocomputing 2: Directions for Research*. Cambridge, Massachusetts: The MIT Press, 1990.

[114] V. Vemuri, *Artificial Neural Networks: Theoretical Concepts*. Washington, D.C.: The IEEE Computer Society Press, 1988.

SUBJECT INDEX